A GUIDE TO WRITING AS AN ENGINEER

A GUIDE TO WRITING AS AN ENGINEER

FOURTH EDITION

David Beer

Department of Electrical and Computer Engineering
University of Texas at Austin

David McMurrey

Formerly of International Business Machines Corporation
Currently, Austin Community College

Publisher: Don Fowley
Acquisitions Editor: Dan Sayre
Editorial Assistant: Jessica Knecht
Senior Product Designer: Jenny Welter
Marketing Manager: Christopher Ruel
Associate Production Manager: Joyce Poh
Production Editor: Jolene Ling
Cover Designer: Kenji Ngieng
Production Management Services: Laserwords Private Limited
Cover Photo Credit: © Rachel Watson/Getty Images, Inc.

This book was set by Laserwords Private Limited.

This book is printed on acid free paper.

Founded in 1807, John Wiley & Sons, Inc. has been a valued source of knowledge and understanding for more than 200 years, helping people around the world meet their needs and fulfill their aspirations. Our company is built on a foundation of principles that include responsibility to the communities we serve and where we live and work. In 2008, we launched a Corporate Citizenship Initiative, a global effort to address the environmental, social, economic, and ethical challenges we face in our business. Among the issues we are addressing are carbon impact, paper specifications and procurement, ethical conduct within our business and among our vendors, and community and charitable support. For more information, please visit our website: www.wiley.com/go/citizenship.

Evaluation copies are provided to qualified academics and professionals for review purposes only, for use in their courses during the next academic year. These copies are licensed and may not be sold or transferred to a third party. Upon completion of the review period, please return the evaluation copy to Wiley. Return instructions and a free of charge return mailing label are available at www.wiley.com/go/returnlabel. If you have chosen to adopt this textbook for use in your course, please accept this book as your complimentary desk copy. Outside of the United States, please contact your local sales representative.

Library of Congress Cataloging-in-Publication Data

Beer, David F.
    A guide to writing as an engineer / David Beer, Department of Electrical and Computer
    Engineering, University of Texas at Austin, David McMurrey, Austin Community
    College.—Fourth edition.
        pages cm
    Includes bibliographical references and index.
    ISBN 978-1-118-30027-5 (pbk.)
1. Technical writing. I. McMurrey, David A. II. Title.
    T11.B396 2014
    808.06'662–dc23
                                                                    2012043890

PREFACE

A Guide to Writing as an Engineer, Fourth Edition, like its previous editions is intended for professional engineers, engineering students, and students in other technical disciplines. The book addresses:

- Important writing concepts that apply to communication in these fields.
- Content, organization, format, and style of various kinds of engineering writing such as reports, proposals, specifications, business letters, and email.
- Oral presentations.
- Methods and resources for finding engineering information, both in traditional ways and online.
- Ethics issues in the field of engineering and strategies for resolving them.
- IEEE citation system for ensuring that the sources of all engineering written work and graphics are properly cited.
- Social media: how professional engineers and engineering students can and are using social media to promote themselves, their organizations, products, and services and take an active contributing role in their profession.

WHAT'S NEW IN THIS EDITION

Here is how we have revised *A Guide to Writing as an Engineer*, Fourth Edition:

- Social media: Once viewed as a fad, social media tools and strategies—such as WordPress blogs, LinkedIn, Twitter, and even Google Plus—have become essential tools for many engineering professionals. Jill Brockmann, of Get-Ace.com, provides us with a practical introduction to these tools in Chapter 12 and specific step-by-step instructions on the companion website.

- Tech boxes: Each chapter contains text boxes that briefly describe exciting innovations and advances in the field of engineering: for example, solar panels integrated with roofing shingles, solar paint, insect cyborg spies equipped with piezoelectric generators, graffiti-resistant surfaces based on scorpion exoskeletons, light-producing bacteria, power-producing kites, pavement tiles that produce electricity when walked on, a device that generates electricity from simple human respiration, and many more.
- New examples: Included are examples involving the University of Maryland Watershed building, winner of the 2011 Solar Decathlon; research on batteries for hybrid vehicles; specifications for the University of Minnesota Centaurus II solar vehicle; Maglev space launch systems; a thermal-release ice-cube maker designed by Carnegie Mellon engineering students.
- Engineering design report: Long overdue, Chapter 6 provides discussion and examples of the engineering design report.
- Writing strategies: Chapter 3 adds strategies for explaining the technical to the nontechnical. Chapter 4 adds strategies for writing in tricky situations.
- Companion website: The website companion for *A Guide to Writing as an Engineer*, Fourth Edition, has been resurrected at www.wiley.com/college/beer. It updates URLs, references, and technical content, as necessary. It now includes interactive quizzes, step-by-step procedures for important software tasks, exercises, additional examples, additional tech box items, and other resources.
- Condensed text: To keep the book trim while adding the chapter on social media, we have reduced the word count in each chapter as much as possible but without harming content.

WHO SHOULD USE THIS BOOK

The idea for this book originally grew from our experience in industry and the engineering communication classroom—in particular, from our wish to write a practical rather than theoretical text that devotes *all* its pages to the communication needs of working engineers and those planning to become engineers. Many engineers and engineering students complain that there is no helpful book on writing aimed specifically for them. Most technical writing texts focus, as their titles imply, on the entire field of technical writing. In other words, they aim to provide total information on everything a technical writer in any profession might be called on to do.

Few engineers have the time to become skilled technical writers, yet all engineers need to know how to communicate effectively. They are required to write numerous short documents and also help put together a variety of much longer ones, but few need acquire the skills of an advanced copy editor, graphic artist, or publisher. For most, engineering is their focus, and although advancement to management might bring considerable increase in communication-related work, these will, for the most part, still be focused on engineering and closely related disciplines.

Thus our purpose in this fourth edition is the same as it has been in previous editions: to write a book that stays close to the real concerns engineers and engineering students have in their everyday working lives. Thus, we give little coverage to some topics focused on at length in traditional textbooks and plenty of coverage to topics that a traditional text might ignore. These choices and priorities reflect what we have found to be important to the audience of this book—engineers and students of technical disciplines.

The book can support writing courses for science and engineering majors, or indeed for any student who wants to write about technology. Teachers will find the exercises at the end of each chapter—as well as in the companion website—good starting points for discussion and homework. The book can also function as a reference and guide for writing and research, documenting research, ethical practice in engineering writing, and making effective oral presentations.

WHAT'S IN THIS BOOK

To keep our book focused squarely on the world of engineering, we have organized the chapters in the following way:

Chapter 1, "Engineers and Writing." Study this chapter if you need to be convinced that writing is important for professional engineers and to find out what they write about.

Chapter 2, "Eliminating Sporadic Noise in Engineering Writing." Study this chapter to learn about and avoid communication problems that distract busy readers, causing momentary annoyances, confusion, distrust, or misunderstanding.

Chapter 3, "Guidelines for Writing Noise-Free Engineering Documents." Use this chapter to learn how to produce effective engineering documents that enable readers to access your information with clarity and ease.

Chapter 4, "Letters, Memoranda, Email, and Other Media for Engineers." Learn format, style, and strategies for office memoranda, business letters, and email. (The survey of alternatives to email such as forums, blogs, and social-networking applications has been moved to the new Chapter 12.)

Chapter 5, "Writing Common Engineering Documents." Study the content, format, and style recommendations for such common engineering documents as inspection and trip reports, laboratory reports, specifications, progress reports, proposals, instructions, and recommendation reports.

Chapter 6, "Writing Research and Design Reports." See a standard format for an engineering report, with special emphasis on content and style for its components. Read guidelines on generating PDFs. New to this book is the discussion and examples of the engineering design report.

Chapter 7, "Constructing Engineering Tables and Graphics." Learn strategies for planning graphics for your reports. Techniques for incorporating illustrations and tables into your technical documents have been moved to the companion website.

Chapter 8, "Accessing Engineering Information." Review strategies on how to plan an information search in traditional libraries as well as in their contemporary online counterparts. See the special section on finding resources available on the Internet.

Chapter 9, "Engineering Your Speaking." Read about strategies for preparing and delivering presentations, either solo or as a team.

Chapter 10, "Writing to Get an Engineering Job." Review strategies for developing application letters and résumés—two of the main tools for getting engineering job. The chapter includes suggestions for engineers just beginning their careers. Information on using social media (such as LinkedIn) for the job search has been moved to the new Chapter 12 on social media.

Chapter 11, "Ethics and Documentation in Engineering Writing." Explore the ethical problems you may encounter and how to resolve them. Use one of the two codes of ethics provided to substantiate your position. Read about plagiarism and review the IEEE system for documenting borrowed information. Sample formats of citations and references are provided.

Chapter 12, "Engineering Your Online Reputation." Design and implement a social media strategy for building an online reputation for yourself, your company or your organization using such tools as WordPress, Facebook, Twitter, LinkedIn, and Google+. Learn how to build a community and curate its contributed information so that that information reliably provides online support for products or services. Put what you learn into practice by using these tools to accomplish one or both of these goals, preferably for a business, organization, product, or service.

ACKNOWLEDGMENTS

Many talented people have played a part, directly or indirectly, in bringing this book to print. We appreciate the input of many students in the Department of Electrical and Computer Engineering at the University of Texas at Austin who are now successfully in industry or graduate school, and we are most grateful to a number of engineering friends at Advanced Micro Devices in Austin.

Also deserving of our gratitude are those professors who assisted us in reviewing the manuscript of earlier editions of this text. Such people include Professor W. Mack Grady, ECE Department, UT Austin; Thomas Ferrara, California State University, Chico; Jon A. Leydens, Colorado School of Mines; Jeanne Lindsell, San Jose State University; Scott Mason, University of Arkansas; Geraldine Milano, New Jersey Institute of Technology; Heather Sheardown, McMaster University; and Marie Zener, Arizona State University.

We especially thank the reviewers of this fourth edition: Elizabeth Hildinger, University of Michigan at Ann Arbor; J. David Baldwin, Oklahoma State University; David Jackson, McMaster University; Michael Polis, Oakland University; and Jay Goldburg, of Marquette University. We also appreciate the help of Clay Spinuzzi of the University of Texas at Austin, Linda M. St. Clair of IBM Corporation Austin; Angelina Lemon of Freescale Semiconductor, Inc.; Susan Ardis, Head Librarian, Engineering Library, UT Austin; Teresa Ashley, reference librarian at Austin Community College; Randy Schrecengost, an Austin-based professional engineer; and Jill Brockmann, Adjunct Associate Professor at Austin Community College and CEO of Get-Ace.com. And of course we sincerely thank our families for the encouragement they have always given us.

CONTENTS

1

ENGINEERS AND WRITING

Poor communication skill is the Achilles' heel of many engineers, both young and experienced—and it can even be a career showstopper. In fact, poor communication skills have probably claimed more casualties than corporate downsizing.

H. T. Roman, "Be a Leader—Mentor Young Engineers,"
IEEE-USA Today's Engineer, November 2002.

It is nearly impossible to overstate the benefits of being able to write well. The importance of the written word in storing, sharing, and communicating ideas at all levels of all organizations makes a poor facility with the mechanics of writing a severely career-limiting fault.

John E. West, *The Only Trait of a Leader: A Field Guide to Success for New Engineers, Scientists, and Technologists*, 2008.

Like a lot of other professionals, many engineers and engineering students dislike writing. After all, don't you go into engineering because you want to work with machines, instruments, and numbers rather than words? Didn't you leave writing behind when you finished English 101? You may have hoped so, but the fact remains—as the above quotes so bluntly indicate—that to be a successful engineer you must be able to write (and speak) effectively. Even if you could set up your own lab in a vacuum and avoid communication with all others, what good would your ideas and discoveries be if they never got beyond your own mind?

If you don't feel you have mastered writing skills, the fault probably is not entirely yours. Few engineering colleges offer adequate (if any) 3 courses in engineering communication, and many students find what writing skills they did possess are badly rusted from lack of use by the time they graduate with an engineering degree.

Ironically, most engineering programs devote less than 5% of their curriculum to communication skills—the very skills that many engineers will use some 20% to 40% of their working time. Even this percentage usually increases with promotion, which is why many young engineers eventually find themselves wishing they had taken more writing courses.

But rather than dwell on the negative, look at the needs and opportunities that exist in engineering writing, and then see how you can best remove barriers to becoming an efficient and effective writer. You'll soon find that the skills you need to write well are no harder to

> **Instant learning?**
>
> Researchers at Boston University and ATR Computational Neuroscience Laboratories in Kyoto, Japan, think that by using decoded neurofeedback, people's brain activity can be trained to match that of someone who possesses a certain skill (for example, writing or piano playing). Don't we wish!
>
> For details, see the Preface for the URL.

acquire than many of the technical skills you have already mastered as an engineer or engineering student. First, here are four factors to consider:

- Engineers write a lot.
- Engineers write many kinds of documents.
- Successful engineering careers require strong writing skills.
- Engineers can learn to write well.

ENGINEERS WRITE A LOT

Many engineers spend over 40% of their work time writing, and usually find the percentage increases as they move up the corporate ladder. It doesn't matter that most of this writing is now sent through email; the need for clear and efficient prose is the same whether it appears on a computer or sheet of paper.

An engineer told us some years ago that while working on the B-1b bomber, he and his colleagues calculated that all the proposals, regulations, manuals, procedures, and memos that the project generated weighed almost as much as the bomber itself. Most large ships carry several tons of maintenance and operations manuals. Two trucks were needed to carry the proposals from Texas to Washington for the ill-fated supercollider project. John Naisbitt estimated in his book *Megatrends* over 25 years ago that some 6,000 to 7,000 scientific articles were being written every day, and even then the amount of recorded scientific and technical information in the world was doubling every five and a half years. Jumping to the present, look what John Bringardner has to say in his short article entitled "Winning the Lawsuit":

> *Way back in the 20th century, when Ford Motor Company was sued over a faulty ignition switch, its lawyers would gird for the discovery process: a labor-intensive ordeal that involved disgorging thousands of pages of company records. These*

days, the number of pages commonly involved in commercial litigation discovery has ballooned into the billions. Attorneys on the hunt for a smoking gun now want to see not just the final engineering plans but the emails, drafts, personal data files, and everything else ever produced in the lead-up to the finished product.

Wired Magazine, July 2008, p. 112.

Who generates and transmits—in print, online, graphically, or orally—all this material, together with countless memos, reports, proposals, manuals, and other technical information? Engineers. Perhaps they get some help from a technical editor if their company employs one, and secretaries may play a part in some cases. Nevertheless, the vast body of technical information available in the world today has its genesis in the writing and speaking of engineers, whether they work alone or in teams. Figure 1-1 shows just one response we got when we randomly asked an engineer friend, who works as a software deployment specialist for a large international company, to outline a typical day at his job (our italics indicate where communication skills are called for).

Friday's Schedule 2/15/08	
7:30	Arrive, *read and reply to several overnight emails.*
8:00	Work on project.
10:30	*Meet with* project manager to *write answer to* department head request.
11:00	*Write up a request* to obtain needed technical support.
11:30	Lunch.
12:00	*Meet with* server group about *submitted application* to fix process problems.
12:20	*Reply to* emails from Sales about prospective customers' *technical questions.*
12:30	*Write to* software vendor about how our product works with their plans.
1:00	*Give presentation to* server hosting group *to explain* what my group is doing.
2:00	*Join* the team *to write up* weekly *progress report.*
2:30	*Write emails to* update customers on the status of solving their problems.
2:45	*Write email reply to* question about *knowledge base article I wrote.*
3:00	*Meet with* group *to discuss project goals* for next four months.
3:30	*Meet with* group to *create presentation of findings* to project management.
4:00	Work on project.
5:00	Leave for day.

Figure 1-1 The working day of a typical engineer calls for plenty of communication skills.

The ability to write effectively is not just a "nice-to-have"; it translates into significant dollars. If the average starting salary for engineers in 2011 is $60,000 and those engineers spend 40% of their time writing, that means they are being paid $24,000 a year to write!

ENGINEERS WRITE MANY KINDS OF DOCUMENTS

As mentioned above, few engineers work in a vacuum. Throughout your career you will interact with a variety of other engineering and non-engineering colleagues, officials, and members of the public. Even if you don't do the actual engineering work, you may have to explain how something was done, should be done, needs to be changed, must be investigated, and so on. The list of all possible engineering situations and contexts in which communication skills are needed is unending. Figure 1-2 identifies just some of the documents you might be involved in producing during your engineering career. (Not all companies label reports by the same name or put them in the same categories as we have.)

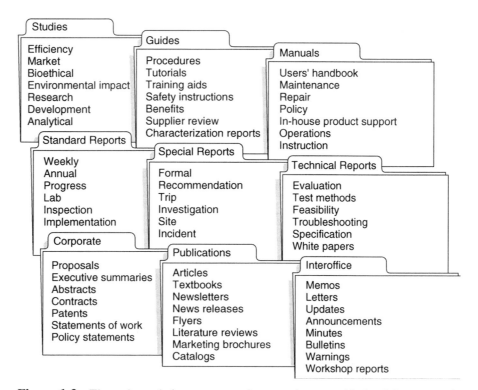

Figure 1-2   Throughout their careers, engineers write many kinds of documents in various contexts and with different purposes and audiences.

Moving further into the twenty-first century, electronic communication is rapidly replacing much hard copy. Used for anything from quick pithy notes and memos to complete multivolume documents, email has perhaps become the most popular form of written communication. Yet this fact does not in any way change the need for clarity and organization in engineering writing, and whatever the future holds, solid skills in clear and efficient writing, and the ability to adapt to many different document specifications, will probably be necessary for as long as humans communicate with each other.

SUCCESSFUL ENGINEERING CAREERS REQUIRE STRONG WRITING SKILLS

In the engineering field, you are rarely judged solely by the quality of your technical expertise or work. People also form opinions of you by what you say and write—and how you say and write it. When you write email or reports, talk to members of a group, deal with vendors on the phone, or attend meetings, the image others get of you is largely formed by how well you communicate. Even if you work for a large company and don't see a lot of high-level managers, those same managers can still gain an impression of you by the quality of your written reports as well as by what your immediate supervisor tells them. Thus Robert W. Lucky, former Executive Director of AT&T Laboratories and head of research at Telcordia Technologies, and an accomplished writer himself, points out:

> *It is unquestionably true that writing and speaking abilities are essential to the successful engineer. Nearly every engineer who has been unsuccessful in my division had poor communication skills. That does not necessarily mean that they failed because of the lack of these skills, but it does provide strong contributory evidence of the need for good communication. On the contrary, I have seen many quite average engineers be successful because of above-average communication skills.*

rlucky@telcordia.com Accessed August 20, 2008

Moreover, two relatively recent trends are now making communication skills even more vital to the engineering profession. These are *specialization* and *accountability*. Due to the advancement and specialization of technology, engineers are finding it increasingly difficult to communicate with one another. Almost daily, engineering fields once considered unified become progressively fragmented, and it's quite possible for two engineers with similar academic degrees to have large knowledge gaps when it comes to each other's work. In practical terms, this means that a fellow engineer may have only a little more understanding of what you are working on than does a layperson. These gaps in knowledge often have to be bridged, but they can't be unless specialists have the skills to communicate clearly and effectively with each other. (Chapter 3 presents "translation" techniques that can help with these gaps as well.)

In addition, because engineers and their companies are now held much more accountable by the public, engineers must also be able to communicate with government, news media, and the general public. As the Director of the Center for Engineering Professionalism at Texas Tech University puts it,

> *The expansiveness of technology is such that now, more than ever, society is holding engineering professionals accountable for decisions that affect a full range of daily life activities. Engineers are now responsible for saying: "Can we do it, should we do it, if we do it, can we control it, and are we willing to be accountable for it?" There have been too many "headline type" instances of technology gone astray for it to be otherwise . . . Pinto automobiles that burn when hit from the rear, DC-10s that crash when cargo doors don't hold, bridges that collapse, Hyatt Regency walkways that fall, space shuttles that explode on national TV, gas leaks that kill thousands, nuclear plant accidents, computer viruses, oil tanker spills, and on and on.*

Engineering Ethics Module, Murdough Center for Engineering Professionalism, Texas Tech University, Lubbock, Texas. www.murdough.ttu.edu/EthicsModule /EthicsModule.htm. Accessed December 13, 2011.

People do want to know *why* a space shuttle crashed (after all, their taxes paid for the mission). They want to know if it really is safe to live near a nuclear reactor or high-power lines. The public—often through the press—wants to know if a plant is environmentally sound or if a project is likely to be worth the tax dollars. Moreover, there is no shortage of lawyers ready to hold engineering firms and projects accountable for their actions. All this means that engineers are being called upon to explain themselves in numerous ways and must now communicate with an increasing variety of people—many of whom are not engineers.

ENGINEERS CAN LEARN TO WRITE WELL

Here are the words of Norman Augustine, former chairman and CEO of Martin Marietta Corporation and also chair of the National Academy of Engineering:

> *Living in a "sound bite" world, engineers must learn to communicate effectively. In my judgment, this remains the greatest shortcoming of most engineers today—particularly insofar as written communication is concerned. It is not sensible to continue to place our candle under a bushel as we too often have in the past. If we put our trust solely in the primacy of logic and technical skills, we will*

lose the contest for the public's attention—and in the end, both the public and the engineer will be the loser.

Norman R. Augustine, in *The Bridge*, The National Academy of Engineering, *24*(3), Fall 1994, p. 13.

Writing is not easy for most of us; it takes practice just like programming, woodworking, or playing the bagpipes, for example. A lot of truth lies in the adage that no one can be a good writer—only a good *re*writer. If you look at the early drafts of the most famous authors' works, you will see scribbling, additions, deletions, rewordings, and corrections where they have edited their text. So don't expect to produce a masterpiece of writing on your first try. Every initial draft of a document, whether it's a one-page memo or a fifty-page set of procedures, needs to be worked on and improved before being sent to its readers.

As an engineer you have been trained to think logically. In the laboratory or workshop, you are concerned with precision and accuracy. From elementary and secondary school, you already possess the skills needed for basic written communication, and every day you are exposed to clear writing in newspapers, weekly news magazines, and popular journal articles. Thus you are already in a good position to become an effective writer partly by emulating what you've already been exposed to. All you need is some instruction and practice. This book will give you plenty of the former, and your engineering career will give you many opportunities for the latter.

NOISE AND THE COMMUNICATION PROCESS

Have you ever been annoyed by someone talking loudly on a cell phone while you were trying to study or talk to a friend? Or maybe you couldn't enjoy your favorite TV show because someone was using the vacuum cleaner in the next room or the stereo was booming.

In each case, what you were experiencing was noise interfering with the transmission of information—specifically, *environmental* noise. In written communication, we are primarily concerned with *syntactic* (grammar), *semantic* (word meanings), and *organizational* noise.

Whenever a message is sent, someone is sending it and someone else is trying to receive it. In communication theory, the sender is the *encoder*, and the receiver is the *decoder*. The message, or *signal*, is sent through a channel, usually speech, writing, or some other conventional set of signs. Anything that prevents the signal from flowing clearly through the channel from the encoder to the decoder is *noise*. Figure 1-3 illustrates this concept. Note how all our actions involving communication are "overshadowed" by the possibility of noise.

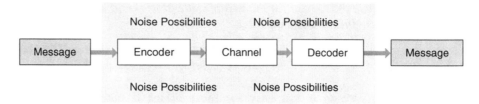

Figure 1-3   In noise-free technical communication, the signal flows from the encoder (writer, speaker) to the decoder (reader, listener) without distortion or ambiguity. When this occurs, the received message is a reliable version of the sent one.

Apply this concept to engineering writing: anything causing a reader to hesitate—whether in uncertainty, frustration, or even unintended amusement—is noise. Chapter 2 will provide more detail, but for now the following box shows just a few simple samples of written noise.

Noisy sentences

When they bought the machine they werent aware of it's shortcomings.

They were under the allusion that the project could be completed in six weeks.

There was not a sufficient enough number of samples to validate the data.

Our intention is to implement the verification of the reliability of the system in the near future.

In the first sentence, two apostrophe problems cause noise. A reader might be distracted momentarily from the sentence's message (or at least waste time wondering about the writer). The same might be said for the confusion between *allusion* and *illusion* in the second sentence. The third sentence is noisy because of the wordiness it contains. Wouldn't you rather just read *There weren't enough samples to validate the data*? The final example is a monument to verbosity. With the noise removed, it simply says: *We want to verify the system's reliability soon*.

It's relatively easy to identify and remove simple noise like this. More challenging is the kind of noise that results from fuzzy and disorganized thinking. Here's a notice posted on a professor's door describing his office hours:

More noise

I open most days about 9 or 9:30, occasionally as early as 8, but some days as late as 10 or 10:30. I close about 4 or 4:30, occasionally around 3:30, but sometimes as late as 6 or 6:30. Sometimes in the mornings or afternoons, I'm not here at all, but lately I've been here just about all the time except when I'm somewhere else, but I should be here then, too.

Academic humor, maybe, but it's not hard to find writing in the engineering world that is equally difficult to interpret, as this excerpt from industrial procedures shows:

Noisy procedure

If containment is not increasing or it is increasing but MG Press is not trending down and PZR level is not decreasing, the Loss of Offsite Power procedure shall be implemented, starting with step 15, unless NAN-S01 and NAN-S02 are de-energized in which case the Reactor Trip procedure shall be performed. But if the containment THRSP is increasing the Excess Steam Demand procedure shall be implemented when MG Press is trending down and the LIOC procedure shall be implemented when the PZR level is decreasing.

Noise in a written document can cause anything from momentary confusion to a complete inability to understand a message. However, noise inevitably costs money—or to put it graphically,

$$\text{NOISE} = \$\$\$\$$$

According to engineer Bill Brennan, a senior member of the technical staff at Advanced Micro Devices (AMD) in Austin, Texas, it costs a minimum of $200 to produce one page of an internal technical report and at least five times that much for one page of a technical conference report. Thus, as you learn to reduce noise in your writing, you will become an increasingly valuable asset to your company.

Noise can also occur in spoken communication, of course, as you will see in Chapter 9. For now, recall how often you've been distracted by a speaker's monotonous tone, nervous cough, clumsy use of notes, or indecipherable graphics—while you just sat there, a captive audience.

The following chapters contain advice, illustrations, and strategies to help you learn to avoid noise in your communication. Try to keep this concept of noise in mind when you write or edit, whether you are working on a five-sentence memo or a 500-page technical manual. Throughout your school years you may have been reprimanded for "poor writing," "mistakes," "errors," "choppy style," and so on. However, as an engineer, think of these problems in terms of *noise to be eliminated from the signal*. For efficient and effective communication to take place, the signal-to-noise ratio must be as high as possible. To put it another way, filter as much noise out of your communication as you can.

CONTROLLING THE WRITING SYSTEM

Engineers frequently design, build, and manage systems made up of interconnected parts. Controls have to be built into such systems to guarantee that they function correctly and reliably and that they produce the desired result. If the ATM chews up your card and spits it back out to you in place of the $200 you had hoped for, you'd claim the system is not working right—or that it is out of control. The system is only functioning reliably if the input (your ATM card) produces the desired output (your $200).

What has this got to do with writing? Consider language as a *system* made up of various components such as sounds, words, clauses, sentences, and so on. Whenever we speak or write, we use this system, and like other systems, it must be controlled if it is to do its job right. The person who supposedly wrote in an accident report, *Coming home, I drove into the wrong house and collided with a tree I didn't have*, was obviously unable to express what really happened. The input (thought) to the system (language) did not have the desired output (meaning) because the writer was not in control of the system or was not thinking clearly.

> ### Brain power
>
> Freer Logic has developed a device called Body Wave that, when coupled with an interactive software package called *Play Attention*, provides interactive feedback and training towards peak mental performance—in particular, the ability to control a computer with your brain.
>
> For details, see the Preface for the URL.

In the same way, an instruction like *Pour the concrete when it is above 40°F* indicates a lack of language control since the writer is not clearly stating whether the concrete or the weather must meet the specification of "above 40°F." Thus you might think of language as a system or even a tool you can learn to control so that it will do exactly what you want it to. Learning to control language, namely to write and speak so you get desired results or feedback, is really not much different than training yourself to operate complex machinery or software systems. You can train yourself to eliminate most, if not all, noise that might occur in your writing and speaking. Figure 1-4 depicts how this works. Note how at the end of the process, your communication often receives "feedback." Feedback—in the form of questions and puzzled looks, for example—gives you an indication of how well you are using the language system.

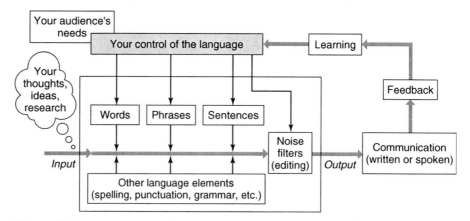

Figure 1-4 The process of communicating can be illustrated as a system with an input and output. How well the input is processed once it is in the system, i.e., how well you convey your information to others, determines the impact of your message. From the response (feedback) you get, you learn how to further improve the process.

If you get the response (or feedback) you want from your communication, you can be pretty sure you have communicated well. A proposal accepted, a repair quickly made, an applied-for promotion awarded—these are just a few examples of the payback from effective communication. To put it another way, if you learn to efficiently control the tool you are using (language) so that it's noise-free, you will produce clear and effective written documents that get results.

EXERCISES

1. Ask any professional engineers about the amount and kinds of writing they do on the job. How much time do they spend writing each day? Is the amount of writing they do related to how long they have been with their company? In what ways have their writing skills helped (or hindered) them in their careers? Do they get any help with their writing from secretaries, peers, or technical writers? What is the attitude of their superiors toward clear writing?

2. Look at the list of technical documents in Figure 1-2. How many are you familiar with? When would they likely be important to you as a reader? Are there other types of documents not included in Figure 1-2? Ask some engineering friends how many kinds of documents they have worked on, either as individuals or as part of a group.

3. Think of your own engineering major or specialty. List some engineering fields related in varying degrees of closeness to yours. What knowledge do you share with people in these fields? What problems can you foresee in communicating with engineers in other fields? What problems would you face if you had to talk about your field to a non-engineering audience?

BIBLIOGRAPHY

Cuevas, Vera. "What Companies Want: The 'Whole Engineer.'" http://business.highbeam.com/3094/article-1G1–21082728/companies-want-whole-engineer. Accessed December 27, 2011.

Ellis, Blake. "Best-paying college major: Engineering." *CNN Money.* http://money.cnn.com/2011/04/08/pf/college/best_paying_college_majors/index.htm. Accessed November 14, 2011.

Miller, Carolyn R. *Communication in the Workplace: A Collaborative Teacher-Student Research Project.* http://www4.ncsu.edu/~crmiller/Publications/ATTW03.pdf. Accessed December 2, 2011.

Naisbitt, John, and Patricia Aburdene. *Megatrends 2000: Ten New Directions for the 1990s.* New York: William & Morrow, 2000.

National Commission on Writing for America's Families, Schools, and Colleges. "Writing Skills Necessary for Employment, Says Big Business." http://www.accountingweb.com/item/99758. Accessed December 2, 2011.

Paradis, James G., and Zimmerman, Muriel L. *The MIT Guide to Science and Engineering Communication,* 2nd ed. Cambridge, MA: MIT Press, 2002.

Rothwell, Dan J. *In the Company of Others: An Introduction to Communication.* New York: McGraw Hill, 2004.

Roy M. Berko, et al., *Communicating: A Social, Career, and Cultural Focus.* 11th ed. Boston: Pearson, 2010.

2

ELIMINATING SPORADIC NOISE IN ENGINEERING WRITING

I am not a picky person when it comes to spelling and grammar, but when I see a report or memo which has repeated errors I immediately question the ability and dedication of the person who wrote it. Why didn't they take the time and effort to do it right? Most of the successful engineers I know write clear, well-organized memos and reports. Engineers who can't write well are definitely held back from career advancement.

Richard L. Levine, Manager, Bell Northern Research, 1987.

There arises from a bad and inapt formation of words, a wonderful obstruction of the mind.

Sir Francis Bacon, 1561–1626.

Errors in writing, causing what Bacon calls "a wonderful obstruction of the mind," are traditionally called faulty mechanics but can be viewed as sporadic or intermittent noise. Enough sporadic noise in a document, such as repeated misspellings or numerous sentence fragments, can easily turn into constant noise. Such noise will give your reader an impression of hastily, carelessly produced work undeserving of the response or feedback you hope for—as is bluntly expressed by an engineering manager in the opening quotation to this chapter.

To help you eliminate intermittent noise, this chapter shows where it is most likely to occur: in spelling, punctuation, sentence structure, and technical usage. This chapter also shows you how to edit your writing in order to remove sporadic noise.

SPELLING AND SPELL CHECKERS

Obviously, electronic spell checkers do not eliminate the need to be a careful speller. With apologies to Shakespeare, we took his words "A rose by any other name would smell as sweet" (from *Romeo and Juliet*) and ran them through a spell checker as *A nose by any outer dame wood small as sweat*. No red flags were raised. Nor will spell checkers catch common errors such as confusing *there* for *their*, *to* for *too*, or *it's* for *its*. Some typographical errors simple give you other words that will pass unnoticed, as in this sentence. (Did you see it?) A very slight slip of the finger on the keyboard can make the difference between asking for some forms to be *mailed* to you or *nailed* to you. A quick transposition could render a memo *nuclear* rather than simply *unclear*.

At best, poor spelling can be annoying to readers, or at least distract them from what you want to communicate. Noise created by misspelling can bring readers to a stop and cause them to seriously question your ability as a writer. They might even suspect that a careless speller could also be inept in more critical technical matters, as the author of the quote at the top of this chapter implies.

To reduce or eliminate any noise in your writing caused by incorrect spelling, use a spell checker but also have a standard dictionary nearby. A current dictionary is the only resource that can reliably answer questions such as the following:

- Whether there is more than one way to spell a word, or what the accepted plural forms of words such as *appendix* or *matrix* are.
- How words like *well-known* or *so-called* are hyphenated, or whether a computer is *on-line* or *online*.
- Whether it is appropriate to write about *FORTRAN, Fortran,* or *fortran*.
- What the difference between British and American spelling or usage might be.
- What the accepted past tense is of recent verbs that have come into technical English such as *input*.

It is especially important for an engineer to use a current dictionary. English is a dynamic language, and the language of science and technology changes even more rapidly as knowledge increases and devices are developed. You won't find words like *software, modem,* and *LED* in a dictionary from the 1950s, and since then older words such as *bug, hardware, interface,* and *mouse* have taken on new meanings. Some usage has yet to be decided on: Would a computer shop advertise that it repairs *mice* or *mouses*? Do you send *e-mail, E-mail,* or *email*? (As of now all three options are still used, but *email* seems to be winning.)

PUNCTUATION

Would you want to drive on a busy highway where there were no traffic signs? Controlling the flow of traffic is vital if anyone is to get anywhere. Similarly, within sentences the flow of meaning is controlled by punctuation marks, the conventionally agreed-upon "traffic signals" of written communication. Spoken language uses an equivalent system: pitch, pauses, and emphasis.

You may want to look at detailed guides to punctuation if you have a lot of queries in this area. You will also find excellent advice on punctuation in standard college dictionaries. And don't forget: September 23 is National Punctuation Day (www.nationalpunctuation-day.com/)! Meanwhile, the following suggestions are offered on the most common problems with punctuation.

> ### Insect power
>
> Insect cyborgs, equipped with batteries, tiny solar cells or piezoelectric generators to harvest energy from the movement of an insect's wings, are being developed as first responders or super stealthy spies. Case Western Reserve engineers have even created a power supply using the insects' normal feeding.
>
> For details, see the Preface for the URL.

COMMAS

There are plenty of stories about comma errors costing millions of dollars. For example, a blog called "A Whole Lot of Nothing" (http://allthingsmundane.wordpress.com/2010/03/23/just-file-it-under-oops-7-costly-clerical-errors) cites the following sentence in which the final comma enabled a supplier company to break the agreement and reap millions of unexpected dollars:

> *[The agreement] shall continue in force for a period of five years from the date it is made, and thereafter for successive five-year terms, unless and until terminated by one year prior notice in writing by either party.*

Why? The comma before the clause at the end of the sentence indicates that that clause is nonrestrictive (covered in the following pages) and therefore not essential to the sentence.

Confusion sometimes exists about commas because in some cases their use is optional. *Before we arrived at the meeting we had already decided how to vote* would be written with a comma after *meeting* by some but not by others. Does adding or omitting a comma in a given sentence create noise, or does it improve clarity? If no possible confusion results, some technical writers omit unessential commas. However, others punctuate according to the structure of the sentence, which is discussed in the following.

Introductory element commas. Often, omitting a comma after introductory words or phrases in a sentence will cause your reader to be momentarily confused—as you would have been if there were no comma after the first word of this sentence. Here are further examples of missing commas causing noise.

Punctuation problem: After the construction workers finished eating rats emerged to look for the scraps.

Revision: After the construction workers finished eating, rats emerged to look for the scraps.

Punctuation problem: Although the CHIP House took about US$1 million to develop producing a duplicate would cost around US$300,000.

Revision: Although the CHIP House took about US$1 million to develop, producing a duplicate would cost around US$300,000.

Punctuation problem: As you can see the efficiency peaks around 10–12%.

Revision: As you can see, the efficiency peaks around 10–12%.

Punctuation problem: If an acoustic horn has a higher throat impedance within a certain frequency range it will act as a filter in that range which is undesirable.

Revision: If an acoustic horn has a higher throat impedance within a certain frequency range, it will act as a filter in that range, which is undesirable.

Try saying these sentences aloud with their intended meanings. You'll find you put the comma—or pause—where it belongs almost without thinking. If you are not sure, just put a comma after the introductory words or phrases—it's never wrong.

Serial commas. Most technical editors prefer to put a comma before the *and* for a list within a sentence: *The serial comma has become practically mandatory in most scientific, technical, and legal writing*. Notice how the serial comma is useful in the following sentences:

Good uses of the serial comma:

Fresnel's equations determine the reflectance, transmittance, phase, and polarization of a light beam at any angle of incidence.

Tomorrow's engineers will have to be able to manage information overload, communicate skillfully, and employ a computer as an extension of themselves.

A serial comma may also prevent confusion:

Potential punctuation problem: Rathjens, Technobuild, Johnson and Turblex build the best turbines for our purposes.

Revision: Rathjens, Technobuild, Johnson, and Turblex build the best turbines for our purposes.

Unless *Johnson and Turblex* is the name of one company, you will need a serial comma.

Commas for restrictive and nonrestrictive elements. Earlier in the section, you read about a single comma costing one company millions and profiting another company millions. The problem involved restrictive and nonrestrictive elements. In the first example below, the "which" clause provides extra, nonessential information about the CHIP House. The second example *restricts* the meaning of "heat" to just that form of heat generated by air conditioning. The "which" clause in the third example provides extra, nonessential, nice-to-know information about the house's insulation.

Restrictive elements:

The car *that has a dented left fender* is mine.

Heat *that is generated by the air conditioning* is used to make hot water.

A net-zero energy home is one *that requires no external energy source.*

Nonrestrictive elements:

My car, *which is a 2012 Ford Focus*, has a dented left fender.

The CHIP House, *which stands for "Compact Hyper-Insulated Prototype,"* was started with the goal of creating a net-zero energy home.

The CHIP House's most striking feature is the insulation fitted around the entire 750-square foot home, *which makes it look like a giant mattress but also preserves the interior temperature.*

Notice in the preceding examples that the nonrestrictive elements typically use *which* and commas, whereas the restrictive clauses use "that" and no commas. (Try going on a "which hunt" and see what you find.)

SEMICOLONS

Like it or not, semicolons seem to be disappearing from engineering writing. Often the semicolon is replaced by a comma, which is an error according to traditional punctuation rules. More frequently we simply use a period and start a new sentence, but then a psychological closeness might be lost. Look at these examples:

Punctuation problem: Your program is working well, however mine is a disaster.

Revision: Your program is working well; however, mine is a disaster.

Punctuation problem: The CHIP House's most striking feature is the insulation fitted around the entire 750-square foot home, this makes it look like a giant mattress but also preserves the interior temperature.

Revision: The CHIP House's most striking feature is the insulation fitted around the entire 750-square foot home; this makes it look like a giant mattress but also preserves the interior temperature.

Semicolons may be disappearing from engineering writing because people feel less confident using them. Perhaps less noise seems to result from using a comma or a period and new sentence, as in the examples above. Note this pair of sentences:

Punctuation problem: The energy efficiency of the CHIP House makes it stand out on its own, however, its smart features move it beyond the typical green-conscious home.

Revision: The energy efficiency of the CHIP House makes it stand out on its own; however, its smart features move it beyond the typical green-conscious home.

If you frequently use words like *however, therefore, namely, consequently*, and *accordingly* to link what could otherwise be two separate sentences, insert a semicolon before and a comma after them. You'll find this will add a shade of meaning that cannot be achieved otherwise.

Use semicolons to separate a series of short statements listed in a sentence if any one of the statements contains internal punctuation. The semicolon will then divide the larger elements:

Semicolon to clarify list elements with their own internal commas:

I suggest you choose one social science course, such as psychology or philosophy; one natural science course, such as chemistry, physics, or biology; and one math course.

The team is made up of Seth Deleery, vice-president of marketing; Nat Beers, director of research; Ruth Ustby, assistant director of training and human relations; and Cate Kanapathy, chief avionics engineer.

COLONS

Other than for time notation and book or article titles, colons are used within sentences to introduce an informal list:

Punctuation problem: For the final exam you will need: a pencil, a calculator, and three sheets of graph paper.

Revision possibilities:

For the final exam you will need several items: a pencil, a calculator, and three sheets of graph paper.

For the final exam you will need a pencil, a calculator, and three sheets of graph paper.

Notice in the problem version that what precedes the colon makes no sense by itself and the colon needlessly interrupts the flow of the sentence. Notice in the revision with the colon that an independent clause—a statement that can stand by itself—comes before the colon.

PARENTHESES

Use parentheses to set off facts or references in your writing—almost like a quick interjection in speech:

Good uses of parentheses:

Resistor R5 introduces feedback in the circuit (see Figure 5).

This reference book (published in 1993) still contains useful information.

If what you place within parentheses is not a complete sentence, put any required comma or period outside the parentheses, as shown in the first and second examples:

Punctuating parenthetical elements:

Typical indoor levels of radon average 1.5 picocuries per liter (a measure of radioactivity per unit volume of air).

Whenever I design a circuit (like this one), I determine the values of the components in advance.

I have already calculated the values of the resistors. (R1 is 10.5 KΩ, and R2 is 98 Ω.) The next step is to choose standard values.

If your parenthetical material forms a complete sentence—as in the third example above—put the period inside the closing parenthesis.

Remember, it is best not to use parenthetical material too frequently since these marks force your readers to pause and are likely to distract them (if only for a brief moment—see what we mean?) from the main intent of your writing.

DASHES

An em dash (the energy efficiency mistakenly referred to as a hyphen) can provide emphasis by calling attention to the words after it: *He was tall, handsome, rich—and stupid*. Since the em dash is considered less formal than the other parenthetical punctuation marks (parentheses and commas), avoid overusing it in very formal writing. With this caution in mind, dashes are helpful for the following purposes:

Emphasis:	Staying up all night to finish a lab project is not so terrible—once in a while.
Summary:	Reading all warnings, wearing safety glasses and hardhats, and avoiding hot materials—all these practices are crucial to sensible workshop procedure.
Insertion:	My opinion—whether you want to hear it or not—is that the drill does not meet the specifications promised by our supplier.

Notice the em dash touches the letters at each end of it. The en dash is shorter, slightly longer than a hyphen, and is used when you cite ranges of numbers: *31–34; $350–400*.

HYPHENS

Hyphens have been called the most underused punctuation marks in technical writing. Omitting them can sometimes create real noise, as when we read *coop* (an enclosure for poultry or rabbits) but discover that *co-op* was meant. Consensus is lacking on whether to hyphenate pairs of words acting as a unit before a noun—as in *The transistor is a twentieth-century invention*. Sometimes a recent dictionary can help, but here are some suggestions:

Insect cyborgs with backpacks

Engineers at University of Michigan are not only using wing movement to harvest energy, but they are equipping insect cyborgs with backpacks loaded with power cameras, microphones, and other sensors.

For details, see the Preface for the URL.

- Don't hyphenate prefixes such as *pre-*, *re-*, *semi-*, *sub-*, and *non-* unless leaving out a hyphen causes possible confusion. *Preconception* is fine, but *preexisting* needs a hyphen if only for looks. The same might be said of *antiinflationary, ultraadaptable,* or *reengineering*. You may have to distinguish, for example, between *recover* (regain) and *re-cover* or *resent* and *re-sent*.

- Don't hyphenate compound words before a noun when the first one ends in *ly*: example, *early warning system, optimally achieved goals, highly sensitive cameras.*
- Stay alert for sentences in which you can eliminate noise by adding one or more hyphens. A hyphen improves the second sentence of each of the following pairs:

Punctuation problem: We used a 16 key keypad.

Revision: We used a 16-key keypad.

Punctuation problem: We knew Marienet made klystrons would be able to generate a 9.395 GHz microwave.

Revision: We knew Marienet-made klystrons would be able to generate a 9.395 GHz microwave.

Punctuation problem: The equation assumes a one dimensional plane wave propagation inside the horn.

Revision: The equation assumes a one-dimensional plane-wave propagation inside the horn.

Punctuation problem: Research showed the computer aided students improved their grades dramatically.

Revision: Research showed the computer-aided students improved their grades dramatically.

But how do you hyphenate really complex technical terms such as *direct axis transient open circuit time constant*? The best solution (*direct-axis transient open-circuit time constant*) may only be found in a technical dictionary or by observing what the common practice is among specialists in the field.

QUOTATION MARKS

Use quotation marks to set off direct quotations in your text, and put any needed period or comma within them, even if the quoted item is only one word. Although British publishers use different guidelines, the American practice is always to put commas and periods inside quotes, and semicolons and colons outside:

Good punctuation of quotations:

The manager stressed to the whole group that the key word was "Preparedness."

"The correct answer is 18.2 Joules," he told me.

We had heard about the "Four-Star Marketing Plan," but no one remembered what it involved.

We left the game right after the band played "The Eyes of Texas"; it was too darned hot and humid to stay any longer.

As for question marks and quotations, if the question mark applies only to what is within the quotes, it goes inside the final quotation marks with no following period. If it applies to the whole sentence, it goes outside the final quotation marks:

Good punctuation of question marks in quotations:

Their manager bluntly asked, "Are we on schedule?"

What is the meaning of the term "antepenultimate"?

If you need to quote material that takes up more than two lines, use a blockquote in which you set it off from your regular text with vertical spaces, indent it from both right and left margin, and omit the quotation marks:

Good use of a block quotation:

According to the author, specifications should not be written by a single person:

> *The lead engineer delegates the writing of numerous sections to specialists, who may not be aware of the overall goals of the project, and may have parochial views about certain requirements. The lead engineer is faced with the difficult task of fitting all these pieces together, finding all the places where they may conflict, and adjusting them to be correct and consistent with each other [NAWCTSD Technical Report 93–022, p.11].*

The importance of consistency cannot be overstressed in the production of . . .

TRADITIONAL SENTENCE ERRORS

Traditional sentence errors are what most of us studied in high school and even in college. Commonly referred to as "grammar" problems, technically these are mostly *usage* problems. Usage refers to the way society uses language, often deeming certain usages as correct only because society has deemed it to be so (for example, *lie* and *lay*).

MAKING SUBJECTS AND VERBS AGREE

It's unlikely you would write *The machines is broken* without quickly noticing a discrepancy between the subject (*machines*) and the verb (*is*). A problem can occur, however, when several words come between your subject and verb and you forget how you started the sentence. If you are writing in a hurry and leave no time for editing, you might produce problems like these:

Agreement problem: This <u>combination</u> of electrical components *constitute* a single-pole RC filter.

Revision: This <u>combination</u> of electrical components *constitutes* a single-pole RC filter.

Agreement problem: A 35 mm <u>film</u> of some high buildings *are* strongly recommended.

Revision: A 35 mm <u>film</u> of some high buildings *is* strongly recommended.

Agreement problem: Only <u>one</u> of the pre-1925 high-rise structures *were* damaged in the quake.

Revision: Only <u>one</u> of the pre-1925 high-rise structures *was* damaged in the quake.

Those plural nouns above (*components, buildings, structures*) are not the true subjects of their sentences. The words preceding them (*combination, film, one*) are. Style and grammar checkers on your word processor are not entirely reliable ways to check for these kinds of errors.

Sometimes a question arises in engineering writing with units of measurement. For example:

Twelve ounces of adhesive (was/were?) added.

<u>Twelve ounces</u> of adhesive *was* added.

Twelve grams of acid (was/were?) spilled.

<u>Twelve grams</u> of acid *was* spilled.

The reason for the singular verbs above is a matter of logic rather than grammar. Even though several ounces or grams are involved, we "see" them as one unit, and thus the singular verb is preferable.

Using *either/or* or *neither/nor* in sentences also creates some special problems as the following examples show:

Either the old manual or the recent procedures (is/are?) acceptable.

Either the old manual or the recent <u>procedures</u> *are* acceptable.

Either the recent procedures or the old manual (is/are?) acceptable.

Either the recent procedures or the old <u>manual</u> *is* acceptable.

The verb agrees with the word following *or* or *nor*. *Neither/nor* works the same way.

MODIFIER PROBLEMS

Another problem that creates noise occurs when modifiers are misplaced in a sentence. A modifier is a word or group of words whose function is to add meaning to other ideas in a sentence. Misplaced modifiers produce sentences that don't make sense or that make sense in the wrong way. For example, readers get the wrong impression (or no impression) about who is doing what in a sentence. This is frequently because words like "I" or "we" or "the engineers" or some other subject has been omitted. Consider the following:

Modifier problem: Jumping briskly into the saddle, the horse galloped across the prairie.

Revision: Jumping briskly into the saddle, the outlaw galloped across the prairie.

Modifier problem: After testing the mechanism, the theory behind it was easily understood.

Revision: After testing the mechanism, we easily understood the theory.

Modifier problem: Once having completed needed modifications and adjustments, the equipment operated correctly and met all specifications.

Revision: Once we had completed needed modifications and adjustments, the equipment operated correctly and met all specifications.

If we look at these problem versions logically, we have a horse that rides, a theory that tests a mechanism, and equipment that modifies and adjusts. In the revisions, notice that the correct subject is put in the main clause (in the first two examples) and in the dependent clause (in the last example).

Meanwhile, another problem can crop up if you place a modifier too far from the word or idea it modifies:

Modifier problem: I was ordered to get there as soon as possible <u>by fax</u>.

Revision: I was ordered <u>by fax</u> to get there as soon as possible.

Modifier problem: By the age of 4, <u>her father</u> knew that *she* would be an engineer.

Revision: By the time <u>his daughter</u> was 4, her father knew that *she* would be an engineer.

It's not hard to remedy the lack of logic in these sentences and to avoid traveling by fax or having 4-year-old fathers, but sometimes the meaning cannot be extracted, as in the following:

Modifier problem: The tone-detector circuit was too unreliable to be used in our telephone answering device, which was built of analog devices.

Revision: The tone-detector circuit, which was built of analog devices, was too unreliable to be used in our telephone answering device.

The sentence would be correct if the telephone answering device was made of analog devices, but much more likely the writer is concerned with the inaccuracies of an analog tone-detector circuit, as shown in the revision.

UNCLEAR PRONOUNS

When you use a pronoun in your writing, it is commonly assumed that you are referring to whatever noun or nouns come just before it in the sentence. Thus, *The promotion was given to Vicky, who really deserved it*, is perfectly clear: The *who* refers to Vicky. Problems can occur, however, especially with the pronouns *this* and *that*, with their plurals, and with *which* and *it*:

Pronoun problem: We will study the terrain by soil analysis and computer simulation before reaching a decision on whether construction can take place here. This will also enable us to . . .

Revision: This study will also enable us to . . .

Pronoun problem: Back in 1954, three researchers made a series of discoveries about the unknown sources of Barbour's early notebooks. These prompted them to further investigate . . .

Revision: These discoveries prompted the three to further investigate . . .

What does the *This* refer to in the second sentence—*study, terrain, analysis, simulation, decision*, or *construction*? It should be *construction* since it's the last noun before the pronoun *This*. However, that's unlikely to be what the writer meant. The meaning is much clearer in the revision. In the second pair of sentences, readers can eventually figure out that *These* refers to *discoveries* and not *sources* or *notebooks*, but we don't want them to have to figure things out.

PARALLELISM

Parallelism refers to items in a list using the same style of phrasing. Faulty parallelism creates noise because it is grammatically inconsistent. Rather than tell someone you *like to jog, wrestling, and play the fiddle*, say that you *like to jog, wrestle, and play the fiddle*, or that you enjoy *jogging, wrestling, and playing the fiddle*. Consider this example:

> *Parallelism problem*: After a lot of discussion, the team concluded that their alternatives were <u>to call</u> in a consultant, thus increasing the cost of the project, or <u>having</u> three more engineers reassigned to the team.
>
> *Revision*: After a lot of discussion, the team concluded that their alternatives were <u>to call</u> in a consultant, thus increasing the cost of the project, or to have three more engineers reassigned to the team.

Note how the problem version reads as if the team's alternatives are (1) to call in a consultant, and (2) having more engineers reassigned—two unparallel phrases that that lack grammatical consistency. The revision states that the alternatives are *to call in a consultant . . . or to have three more engineers reassigned*. See if you can recognize the lack of parallelism in the problem version:

> *Parallelism problem*: The back-up system should be efficient, should meet safety specifications, and have complete reliability.
>
> *Revision possibilities*:
>
> The back-up system <u>should</u> be efficient, <u>should</u> meet safety specifications, and <u>should</u> be completely reliable.
>
> The back-up system should be efficient, meet safety specifications, and be completely reliable.

Keeping parallel structure is even more important when you construct lists, as Chapter 3 will show.

FRAGMENTS

Sentence fragments are partial statements that create noise because they convey an incomplete idea. Here's an example:

Fragment: She decided to major in petroleum engineering. Even though it would take five years.

Revision: She decided to major in petroleum engineering even though it would take five years.

In the problem version, the first sentence makes sense by itself. Try saying the second statement alone, independent from the first, and your listeners will be lost. True, in everyday speech and popular journalism you will find plenty of fragments that seem to cause little or no noise:

Nonproblem fragment: The *Kinectimals* video game lets players pet a virtual pet on their TV screen. But not actually groom their pets remotely!

Revision: The *Kinectimals* video game lets players pet a virtual pet on their TV screen—but not actually groom their pets remotely!

The nonproblem fragment above is fun in a popular, journalistic context, but it could not stand alone and make sense. In your formal engineering writing you would do well to avoid fragments. They can usually be quite easily remedied, as you can see.

> **Cyborg insect warriors**
>
> Researchers at DARPA are looking at ways to enable remote-controlled dragon-flies to transmit video and other environmental data from the battlefield frontlines.
>
> For details, see the Preface for the URL.

TWO LATIN LEGACIES

Taught to us in the past, a few grammar rules do not hold up under careful linguistic or logical inspection. They were based on how Latin works, rather than English. To put it another way, noise seldom occurs when these rules are ignored. Here are the two main ones, together with comments and a caution.

"Never End a Sentence with a Preposition."   One of the strange taboos is not to end a sentence (or in fact any clause) with a preposition. In reality, that is often the best word to end a sentence with. (A purist might claim we should have just written...*the best word with which to end a sentence*). When an editor criticized Sir Winston Churchill for doing so, Churchill responded with "Young man, this is the

kind of nonsense up with which I will not put!" Did you even notice that we broke the "rule" in the second sentence? Efficient writing sometimes dictates that we end a sentence with a preposition. Compare the following pairs of example. You can see that in each case the second natural version, ending with a preposition, flows better and is more natural:

Hypercorrect version: That's a problem on which we will really have to work.

Natural version: That's a problem that we will really have to work on.

Hypercorrect version: We must make sure we can find some engineering consultants on whom we can really count.

Natural version: We must make sure we can find some engineering consultants we can really count on.

"Never Split an Infinitive." An infinitive is the form of a verb combined with the word *to*, as in *to go, to work*, or *to think*. Confident writers have dared *to deliberately split* the infinitive whenever doing so was in the best interests of clear writing. For a long time now, certain TV space adventurers have been venturing *to boldly go* where the rest of us can't. Sometimes, an electrician may find it necessary (and safer) *to entirely separate* the wires in a power line. But don't overload a split infinitive by putting too many words between *to* and the rest of the verb:

Split infinitive: The team has been unable to, <u>except for the lead engineer and one technician who is on temporary assignment with us,</u> master the new program.

Revision possibilities:

Except for the lead engineer and one technician on temporary assignment with us, the team has been unable <u>to master</u> the new program.

The team has been unable <u>to master</u> the new program—with the exception of the lead engineer and one technician who is on temporary assignment with us.

SEXIST LANGUAGE

Gender, or sex, is now only indicated in English by *she/he, his/hers, her/him*, and by a small group of words describing activities formerly pursued by one sex or the other, such as *mailman, stewardess, chairman*, or *seamstress*. Now of course men might bring the drinks on an airplane and women might deliver the mail, not to mention take an equal place in the engineering workplace. Given this situation, it is unnecessarily restrictive—and to some people offensive—to use gender-specific terms in writing and speech. In the following pairs, the problem versions are restrictive; the revised versions are inclusive. The revisions show how you can easily reword your sentences to include everyone they should:

Sexist language: Every engineer should be at his workstation by 9 a.m.

Revision possibilities:

Every engineer should be at his or her workstation by 9 a.m.

Engineers should be at their workstations by 9 a.m.

Sexist language: An employee can expect a lot of challenges during his career here.

Revision: Employees can expect a lot of challenges during their careers here.

Sexist language: Every technician must wear safety glasses when he enters the work area.

Revision: Technicians must wear safety glasses when entering the work area.

Most nouns indicating gender in English have already been modified to be inclusive. One title that still sneaks through, however, especially in organizations traditionally dominated by males, is *chairman*. If the "chairman" is female, is she the chairwoman or chairperson? Both are acceptable, but it's probably simpler to refer to anyone in such a position as *chair*:

Sexist language: Sarah is chairman of the new committee on marketing strategy.

Revision possibilities:

Sarah is chair of the new committee on marketing strategy.

Sarah is chairing the new committee on marketing strategy.

SENTENCE LENGTH

When dealing with highly technical subjects, you should rarely write sentences over 20 words long. Technical material can be difficult enough as it is. This difficulty increases if your audience is less familiar with your field than you are. Even nontechnical ideas are hard to grasp in long-winded sentences:

Overly long sentence: We finally had a long discussion with the R & D staff but were not able to convince them that they should commit to a specific date for implementation of the design, but instead they responded with a proposal to extend the project, which would result in a lot more work for all of us and a considerable loss of profits for the company.

Revision: We finally had a long discussion with the R & D staff about the implementation of the design. However, we were not able to convince them that they should commit to a specific date. Instead, they responded with a proposal to extend the project. Unfortunately, that would result in a lot more work for all of us and a considerable loss of profits for the company.

Nobody wants to be left breathless at the end of a mile-long sentence. If you find your sentences tend to be lengthy, look for ways to break them into two or more separate ones. The readability of your prose will be determined partly by the length of your sentences. On the other hand, too many short sentences may leave your readers feeling like first graders:

Short, choppy sentences: The Kw766XTR is a low-profile desktop scanner. It has outstanding performance. It offers a frequency range of 29–54 and 108–174 MHz. It includes 50 memory channels. The design is sleek. Individual channels can be locked out. They can also be delayed.

Revision: The Kw766XTR, a sleek, low-profile desktop scanner, and has outstanding performance, offering a frequency range of 29–54 and 108–174 MHz with 50 memory channels. Individual channels can be locked out or delayed.

TECHNICAL USAGE

Technical usage involves the technical level of the words you use, abbreviations, numbers, units of measurement, acronyms, and equations—the mechanics of your written work.

USELESS AND USEFUL JARGON

"Jargon" is such a pejorative term that you might think that it is a bad thing and nothing else. Read on!

Useless Jargon.   In its negative sense, jargon is pure noise since it refers to unintelligible speech or writing. The word derives from a French verb meaning the twittering of birds, and has a lot in common with "gobbledygook," first used to compare the speech of Washington politicians to the gobbling of Texas turkeys. High-tech jargon is sometimes known as technobabble or scispeak. You see it in such impressive-sounding

	Column 1	*Column 2*	*Column 3*
0.	voltaic	integrated	simulation
1.	Sholokhov's	semiconductor	algorithm
2.	differential	Yagi	attenuator
5.	virtual	tracking	parameters
3.	Fourier	scaled	emission
4.	transient	Q-factor	diode
6.	phasor	diffusion	network
7.	compound	Doppler	gate
8.	thermal	heterodyne	transducer
9.	Gaussian	coaxial	magnetron

Figure 2-1 The Electrotechnophrase Generator (courtesy of ECE students at the University of Texas at Austin). Warning: Use only when you want to be sure no one understands you.

phrases as *integrated logistical programming, differential heterodyne emission*, or *functional cognitive parameters.* Unless these words hold a precise meaning for both writer and reader, no communication takes place—only noise.

Technobabble is so common that with tongue-in-cheek we have provided an "electrotechnophrase generator" in Figure 2-1 to help addicts satisfy their habit. Select any three-digit number and read off the corresponding words from the chart below; for example, 2-8-3 gets *differential heterodyne emission.* Readers may have no idea what you mean, but they should be impressed.

Scorpions against grafitti

Chemical engineers at Jilin University, PRC, are studying the exoskeleton of the yellow fattail scorpion to find out how to create wear-resistant, grafitti-resistant surfaces.

For details, see the Preface for the URL.

Useful Jargon. In another sense, however, jargon is the necessary technical terminology used in specialized fields. A chemical engineer might use the term *deoxyribose* around a group of peers without needing to explain it, just as a geologist could talk about the *Paleozoic* era or *Devonian* period with other geologists. Computer engineers can safely refer to *bytes, bauds,* and *packet switching*—among themselves. Experts need specialized jargon.

As an engineer, you know and use your technical jargon. Some you share with practically all engineers, some with those in the same field of engineering as you—such as chemical, civil, or aerospace—and some you would use only among peers in your highly specialized fields like software engineering.

The way to avoid noise when using technical terminology is to *know your audience.* Make certain you are writing or speaking at their level of comprehension; if you're above their heads, you are wasting your time and theirs. Explain terms whenever necessary; don't risk confusing readers or completely losing them because they don't know what you are talking about. Definitions within your text, examples, analogies, or a good glossary are all useful tools for communicating with less technical audiences.

ABBREVIATIONS

Abbreviations are necessary in technical communication for the same reason that useful technical jargon is: They refer to concepts that would take too much space to spell out fully every time. It would be time-consuming and boring for a computer expert to read *Computer-Aided Design/Computer-Aided Manufacturing* multiple times when *CAD/CAM* would do. However, you will create a lot of noise in your writing if you use abbreviations your readers don't understand. Always spell abbreviations out the first time you use them unless you know this would insult the intelligence of your audience:

Good uses of acronyms:

Then it goes into read-only memory (ROM).

To understand our billing process, you first need to know what a British Thermal Unit (BTU) is.

Once you have defined an abbreviation, you can expect your reader to remember it. Don't use initial caps on the spelled-out version of the abbreviation. Be aware that unnecessary capitalization creates noise—distractions that readers would prefer to skip. Notice above that while it's *ROM*, the spelled-out verson is *read-only memory*.

Some people make a distinction between *initialisms* and *acronyms*. Initialisms take the first letters from each word and pronounce them using those initials: *GPA, IBM, LED, UHF*. Acronyms do the same but are pronounced as words: *AIDS, FORTRAN, NAFTA, NASA, RAM, ROM*. Some acronyms become so commonplace that we think of them as ordinary words and write them in lower case: *bit, laser, pixel, radar, scuba, sonar*.

Two usage pointers:

1. Use the correct form of *a/an* before an initialism. No matter what the first letter is, if it is pronounced with an initial vowel sound (for example, the letter M is pronounced "em"), write *an* before it:

---

an MTCR (Missile Technology Control Regime)

an LED readout

an SRU pin

an ultrasonic frequency (but *a UHF receiver*)

Some abbreviations might fool you. Consider LEM (lunar excursion module) for example. If the custom is to pronounce it as an initialism, L-E-M, then you will have *an* LEM.

2. Form the plural of acronyms and initializations by adding a lowercase *s*. Only put an apostrophe between the abbreviation and the *s* if you are indicating a possessive form:

We ordered three CRTs.

We weren't satisfied with the last CD-ROM's performance.

or

We weren't satisfied with the performance on the last CD-ROM.

NUMBERS

Engineering means working with numbers a great deal. Here, much written noise can occur due to typos, incorrect or inexact numbers, and inconsistencies. Obviously, you can avoid serious noise by making certain any number you write is accurate. Also, give numbers to the necessary degree of precision: Know whether 54.18543 is needed in your report or whether 54.2 will do. Numbers are expressed as words (twelve) or numerals (12). Cardinal numbers are *one, two, three*, etc. Ordinal numbers are *first, second, third*, etc.

Avoid noise from inconsistent use of numbers by following these guidelines:

1. In mainstream nontechnical text, the rule is to write the cardinal numbers from one to ten as words and all other numbers as figures.

---

two transistors   232 stainless steel bolts
three linear actuators 12 capacitors

---

2. However, if the numeric value represents a critical value, and you are writing in a technical context, use the numeral. If the numeric value is not critical, use the word:

Make sure that the rim of the basket is exactly 10 feet from the floor.
She presented three arguments to support her assertion.

---

3. Use words, not numbers, when expressing rounded or approximate numbers:

There are now thousands of apps for the iPhone.
On Wikipedia, there are over three million articles in English.
or
On Wikipedia, there are over 3 million articles in English.

---

4. When more than one number appears in a sentence, write them all the same:

The IPET has 4000 members and 134 chapters in 6 regions.

5. If two numeric values occur in a sequence, make one of them a word and the other numerals:

> The project calls for eight 6-foot boards.

6. Spell out ordinal numbers only if they are single words. Write the rest as numerals plus the last two letters of the ordinal:

> second harmonic   21st element
> fourteenth attempt   73rd cycle

7. If a number begins a sentence, it's a good idea to spell it out regardless of any other rule:

> Thirty-two computers were manufactured today.

8. To avoid writing out a large number at the beginning of a sentence, rewrite the sentence so it doesn't begin with a number:

> Last year, 5198 engines were manufactured in this division.
> *or*
> This division manufactured 5198 engines last year.

9. Punctuate large numbers according to your company's preference or that of your audience. Some countries use periods as decimal markers. Thus 10,354,978 and 10 354 978 and 10.354.978 can all mean different things in different parts of the world.

10. Form the plural of a numeral by adding an *s*, with no apostrophe. Make a written number plural by adding *s*, *es*, or by dropping the *y* and adding *ies*:

> 80s   1920s
> nines   sixes
> fours   nineties

11. Place a zero before the decimal point for numbers less than one. Omit all trailing zeros unless they are needed to indicate precision.

> 0.345 cm   12.00 ft
> 0.5 A   19.40 tons

12. Write fractions as numerals when they are joined by a whole number. Connect the whole number and the fraction by a hyphen:

2-1/2 liters   32-2/3 km

13. Time can be written out when not followed by a.m. or p.m. Use numerals to express time in hours and minutes when followed by a.m. and p.m. or when recording data. Universal Time (UTC, from the French for *universal coordinated time*) uses the 24-hour clock.

ten o'clock   10:41 a.m.
8:45 p.m.   4 hours 36 minutes
12 seconds   23:41 (= 11:41 p.m.)

14. When expressing very large or small numbers, use scientific notation. Some numbers are easily read when expressed in either standard or scientific form. Choose the best format and be consistent:

0.0538 m   *or* 5.38×10^{-2} m
8.32×10^{-21} m/s   *or*   367 345 199 m/s

UNITS OF MEASUREMENT

Although the U.S. public is not committed to the metric system, the engineering profession is. Two versions of the metric system exist, but the more modern one, the SI (from French *Système International*), is preferred. The vital rule is not to mix English and metric units unless you are forced to. If you do not write out the complete word, be sure to use the commonly accepted abbreviation or symbol for a unit, and leave a space between the numeral and the unit.

70 ns	100 dB
12 V	34.62 m
23 e/cm^3	6 Wb/m^2

To accommodate people who still think in English units of measurement, provide both versions in your writing. As with many other editorial matters, make this decision after thinking of your readers' needs. When it might be advisable to add "explanatory" units, as with a mixed audience, do so by writing them in parentheses after the primary units:

212°F (100°C)   5.08 cm (2 in)

Make sure you use the correct symbol when referring to units of measurement, and remember that similar symbols may stand for more than one thing. A great deal of noise (or disaster) could result if you confused the following, for example:

°C (degrees Celsius)	C (coulomb — unit of electric charge)
g (gram)	G (gauss — measure of magnetic induction)
m (thousandth)	M (million)
n (nano-)	N newtons
s (second — as in time)	S (siemens — unit of conductance)

Units of measurement derived from a person's name usually are not capitalized, even if the abbreviation for the unit is. Note that when the name can take a plural form, an *s* is not added to the abbreviation (2048 KB equals 2 MB).

amperes A	farads F	henrys H	kelvins K
teslas T	volts V	webers Wb	

When working with very large or very small units of measurement, be familiar with the designated SI expressions and prefixes:

Factor	Prefix	Symbol
10^{24}	yotta	Y
10^{21}	zetta	Z
10^{18}	exa-	E
10^{15}	peta-	P
10^{12}	tera-	T
10^{9}	giga-	G
10^{6}	mega-	M
10^{3}	kilo-	k
10^{2}	hecto-	h
10^{1}	deka-	da
10^{-1}	deci-	d
10^{-2}	centi-	c
10^{-3}	milli-	m
10^{-6}	micro-	μ
10^{-9}	nano-	n
10^{-12}	pico-	p
10^{-15}	femto-	f
10^{-18}	atto-	a
10^{-21}	zepto	z
10^{-24}	yocto	y

Use a recent dictionary of scientific terms if you are unsure of the correct spellings or symbols of the units you are using. Symbols and abbreviations are indispensable to an engineer, but use them sparingly when writing for an audience other than your peers. Sometimes, you can define things parenthetically or with annotations, as in the following example:

$$P = IE \tag{1}$$

where
$P =$ power, measured in watts
$I =$ current in amperes
$E =$ EMF (electromotive force) in volts

EQUATIONS

It would be hard to do much engineering without equations. They can communicate ideas far more efficiently than words can—consider the ideas represented by $E = mc^2$ for example.

Many word-processing programs now make it easy to write equations in text. Whether you use the computer to write equations or write them in longhand, take care to ensure accuracy and legibility. An illegible or ambiguous equation is hardly going to communicate data effectively, and an error in an equation could be fatal. In other words, make sure your equations are noise-free.

Normally, center equations and number them sequentially in parentheses to the right for reference. See Figure 2-2. Leave a space between your text and any equation, and between lines of equations.

> **Bacteria light & power**
>
> Engineers at the Dutch corporation Philips are studying how fireflies and deep-sea organisms create green light powered by glowing bioluminescent bacteria.
>
> For details, see the Preface for the URL.

Also, space on both sides of operators such as $=$, $+$, or $-$, as shown in the equations on the following page. If you have more than one equation in your document, keep the equal signs and reference numbers parallel throughout:

$$F(x) = \int \log x \, dx \tag{1}$$

$$H(s)(xv_{2)} = X(s)/Y(s) \tag{2}$$

The total harmonic distortion (THD) of voltage at any bus k is defined as

$$THD_k = \frac{\sqrt{\sum_{h=2}^{H} |V_k^h|^2}}{|V_k^l|}, \tag{3}$$

THD can be incorporated into the minimization procedure in [2] by considering a network function that equals the sum of squared THD$_{ks}$, or

$$f(I_m) = \sum_{k-1}^{K} (THD_k)^2 = \sum_{k-1}^{K} \left[\frac{\sqrt{\sum_{h=2}^{H} |V_k^h|^2}}{|V_k^l|} \right]^2$$

$$= \sum_{h=2}^{H} \sum_{k=1}^{K} \frac{1}{|V_k^l|^2} |V_k^h|^2.$$

Note that (2-2) is identical to (2) when y(h) = 1 for h = 2, 3, 4, ..., H, and when

$$b(k) = \frac{1}{|V_K^l|^2}, \quad k = 1, 2, 3, \ldots, K. \tag{4}$$

Since the fundamental frequency voltages are approximately 1.0 pu, the objective function of (2-2) is a close approximation to that of (1).

Figure 2-2 Examples of multiline equations. Notice the standard punctuation marks at the appropriate places.

EDIT, EDIT, EDIT

If you look at the early handwritten drafts of some of the greatest writers' works, you'll see alterations, additions, deletions, and other squiggles that indicate how much revision went into the draft before it became a finished work. We could all produce better written documents if we always:

1. *had* the time to edit our work carefully.
2. *took* the trouble to edit our work carefully.

For an engineer, time is frequently a problem. You can't always find time for a leisurely edit of your work. However, you would be ill-advised to send a first draft of anything of importance to your readers. You must look over anything with an editorial eye, especially if it's going beyond your immediate colleagues. How much time you invest in editing should be in direct proportion to the importance of the document. Use all the assistance your word processor will give you, including any spelling, grammar, or readability programs you may have, but don't follow their suggestions blindly. *You* have to be the final arbiter on the clarity and effectiveness of your work—*your* name will be on the document, not your word processor's manufacturer.

COLLABORATIVE PROOFREADING

Nothing is wrong with having someone look over your writing before you submit it to its intended audience. Two heads are usually better than one for discovering flaws in writing. In industry, experts often cooperate in writing reports, proposals, and other documents just as they work together on engineering projects. In fact, most lengthy documents are produced by team effort, where different team members use their particular strengths.

Collaborative editing can be as simple as asking a friend for his or her opinion of your work and using those comments to improve your writing. The more skilled and frank your friend is, the better. With a long document, however, collaborative editing can be done by having different team members check the document for different potential kinds of noise. This team approach to editing is fully discussed at the end of Chapter 3, under "Share the Load: Write as a Team." Chapter 3 also gives you several guidelines on how to eliminate noise not just within a sentence but in larger chunks of writing—or even throughout an entire document.

EXERCISES

1. Review some of your writing for problems with spelling, punctuation, or the problems discussed in "Sentence Sense." Did you create any noise in your documents by not following these guidelines? How could you use the guidelines as a "quality control" tool when writing in the future?

2. Carefully analyze what you think is good engineering writing in any field. What makes it effective, noise-free writing? List and give examples of the ways in which the writer has carefully observed many of the guidelines given in this chapter.

3. Look at an article in a professional journal or the first chapter of a textbook and determine who its assumed audience is. Then investigate how the author uses technical terminology. Is it appropriate for the audience? Are explanations or definitions given where they seem called for? Do you find any examples of unnecessary technical jargon?

4. In technical journals, review how reports or articles use abbreviations, numbers, units of measurement, and equations. Are they consistent internally? Do they vary between reports and journals? In the case of journals, does the journal provide a style guide for contributing writers?

BIBLIOGRAPHY

Hale, Constance, ed. *Wired Style: Principles of English Usage in the Digital Age*, rev. ed. San Francisco: HardWired, 1999.

Pearsall, Thomas E. *The Elements of Technical Writing*, 2nd ed. Boston: Allyn & Bacon, 2001.

Rude, Carolyn and Angela Eaton. *Technical Editing*, 5th ed. New York: Longman, 2010.

Strunk, William. *The Elements of Style: The Original Edition*. New York: Dover, 2006.

The Chicago Manual of Style, 16th ed. Chicago: The University of Chicago Press, 2010.

3

GUIDELINES FOR WRITING NOISE-FREE ENGINEERING DOCUMENTS

As every engineer knows, form and content must work together. What is sometimes forgotten is that the relationship of form and content applies to documents as well as to physical phenomena. Without some type of form, be it well or poorly structured, no content can be communicated. . . . Even the word "in-form-ation" implies that ideas must be structured in some fashion or other.

Susan Stevenson and Steve Whitmore, *Strategies for Engineering Communication* (New York: John Wiley & Sons, 2002), p. 247.

Information isn't a scarce commodity, as a leading economist wrote in the 1970s. Attention is. So, what can you do to sustain your reader's attention?

Bruce Ross-Larson, *Writing for the Information Age* (New York, W. W. Norton & Company, 2002), p. 3.0.

This chapter presents guidelines for producing large sections of noise-free writing, from efficient paragraphs to effective and useful documents. These guidelines follow the overall process used by successful engineering writers and include important considerations for your entire writing process. The topics covered here represent common problems you as an engineer are likely to face in the course of writing and formatting your documents.

FOCUS ON WHY YOU ARE WRITING

Consider this statement by Ruth Savakinas:

> *Complex technical writing is likely to be very difficult to read. Readability further decreases when the writer does not define major ideas for the reader and when the written document is not relevant to the reader's experience and interests. These two impediments can be eliminated if you clearly define your purpose and your audience. . . .*

From "Ready, Aim—Write!" *IEEE Transactions in Professional Communication*, *31*(1), March 1988, p. 5.

What she wrote over twenty years ago still holds: Before starting to write, have a good idea of precisely who your audience is and what you want to communicate to them. If these goals aren't first defined, your readers may not get a clear message. Thus, whether you have to write a short memo or a lengthy technical report, you should start with a firm sense of purpose so you can (1) present appropriate supporting data, (2) test its adequacy, and (3) discard anything that is not needed.

> ### Kite power
>
> Kites may soon gather energy in high altitudes (where winds are stronger and more consistent) in order to rotate a ground-based turbine. 500 GWh a year?
>
> For details, see the Preface for the URL.

Broadly speaking, the purpose of most technical writing is either to present information or to persuade people to act or think in a certain way. Frequently, however, your documents will have to be both informative and persuasive. To fine-tune your sense of purpose before writing, ask yourself the following:

Do I want to

1. **Inform**—provide information without necessarily expecting any action?
2. **Request**—obtain permission, information, approval, help, or funding?
3. **Instruct**—give information in the form of directions, instructions, or procedures so that my readers will be able to do something?
4. **Propose**—suggest a plan of action or respond to a request for a proposal?
5. **Recommend**—suggest an action or series of actions based on alternative possibilities that I've evaluated?
6. **Persuade**—convince or "sell" my readers, or change their behavior or attitudes based on what I believe to be valid opinion or evidence?
7. **Record**—document for the record how something was researched, carried out, tested, altered, or repaired?

How you write any document should be guided by what you want your audience to do with your information and what they need from the document in order to be able to do it. Thus, your audience plays a defining role in determining how you approach your task. Only a careful analysis of your purpose (or purposes) for writing and the nature of your audience can give you the answers and thus enable you to write to the point.

FOCUS ON YOUR READERS

If you found yourself in a remote region and met people who had never seen anything electronic, you wouldn't hand them your scientific calculator or iPhone and expect them to use it. First, a great deal of technology transfer would have to take place. This may seem obvious, but a lot of technical writing fails because writers make inaccurate assumptions regarding their audience. Engineers often write without taking adequate time initially to analyze those who must read their work. Since you write for many different audiences during your career, as Figure 3-1 shows, take the time to analyze your audience before writing to them.

No matter who you write to, you write because you expect some kind of resulting action—even if it is only nonphysical "action" such as permission, understanding, or a change of opinion. To get results, your communication must bridge a gap between you and your target audience, likely to involve *knowledge*, *ability*, or *interest*. To determine where you stand before writing, first identify who your audience is and then ask yourself these questions:

Figure 3-1   You will deal with many different people as your career progresses, so it is best to have a clear picture of who your audience is before beginning to communicate with them.

Knowledge

- Are my readers engineers in my field who are seeking technical information? Will they be offended or bored by elementary details?
- Are they engineers from a different field who will need some general technical background first?
- Are they managers or supervisors who may be less knowledgeable in my field but who need to make executive decisions based on what I write?
- Are they technicians or others without my expertise and training but with a strong practical knowledge of the field?
- Are they nonexperts from marketing, sales, finance, or other fields who lack engineering or technical backgrounds but who are interested in the subject for nonengineering reasons?
- Are they a mixed audience, such as a panel or committee, made up of experts and lay people?

Ability

- Am I communicating technical information on a level my audience can use?
- Am I using appropriate vocabulary, examples, definitions, and depth of detail?
- Am I expecting more expertise, skill, or action from my audience than I reasonably should?

Interest

- Why will my audience want to spend time reading this document?
- Does my document provide the right level of detail and technology to keep my audience's interest without losing them or boring them?
- What is their current attitude likely to be—positive, neutral, or negative?
- Will my document give them the information they want?

SATISFY DOCUMENT SPECIFICATIONS

Before writing, you should be aware of any specifications your document must meet. If management asks for a brief memo, they may be irritated when you overload their circuits with a lengthy, detailed treatise. When a technician requests the specs on a frequency tester, it won't be appreciated if you write about the strengths and weaknesses of the equipment. If you respond to an RFP (request for proposal) that calls for a proposal of no more than ten pages but submit something twice that long, your proposal will likely be eliminated from the competition.

Many documents have *specifications*. Such specifications may require you to provide sections such as experimental problems, environmental impact, decisions reached, budget estimates, and so on. Specification for an engineering journal may limit the number of words and the number of graphics your technical paper can include. Specification may limit the length of an abstract or summary. Here are the specifications for a research grant proposal:

> Also required is a nontechnical summary (250 words or less) of the research proposed, expressing significance attached to the project and reasons for undertaking it. This summary will be used for public information and should be written in terms that nonscientists can easily understand. .

Many reports have specifications not only for their length but also for such matters as headings, spacing, and margin width. Some government agencies, for example, require that the proposals they receive be written in specific formats, in certain fonts, and even with restrictions on how many letters are permitted in each line of text. Here is an example from an RFP for a government research program:

> Each proposal shall consist of not more than five single-spaced pages plus a cover page, a budget page, a summary page of no more than 300 words, and a page detailing current research funding. All text shall be printed in single-column format on 8-1/2 × 11-inch paper with margins of at least 1 inch on all sides. . . .

GET TO THE POINT

Anyone reading your memos, letters, and reports is likely to be in a hurry. Few engineers have the leisure for "biblical" reading—where one reads from Genesis to Revelation to discover how things turn out. Your documents need to have the most important information at the beginning. Start with your key points, conclusions, or recommendations before presenting the supporting details. For instance, if you test to determine whether some equipment should be replaced, your supervisor will want to know what you have found out and what you recommend. A complete, detailed description of your test procedures may be necessary to support your main points and will likely be verified by others—but it could go unread by those people in management who need only the bottom line.

Where you tell your readers what they most need to know depends on the kind of document you're writing. In a letter, it will be in the opening sentences just after an indication of the purpose of the letter. In a memo, you provide a subject line with a specific reference to the topic. Consider these examples:

> *Vague*: SUBJECT: Employee safety
> *Better*: SUBJECT: Need for employees to wear hard hats and safety glasses
> *Vague*: SUBJECT: Emergency requisitions
> *Better*: SUBJECT: Recommendations to change the procedures for making emergency requisitions

In a longer report, your main points should become quickly evident to your reader through an informative title followed by a summary or abstract of your findings, conclusions, recommendations, or whatever the important information is. (See the chapters on individual reports and the sections on abstracts and executive summaries.) No matter what kind of document you are producing, however, place key information where readers can most efficiently access it.

PROVIDE ACCURATE INFORMATION

Even the clearest writing is useless when the information it conveys is wrong. If you state that an ampere is defined as a coulomb of charges passing a given point in 10 seconds rather than 1 second, you have presented wrong information. If you refer to data in Appendix B of your report when you mean Appendix D, the error could stump your readers and cause them to lose confidence in your report.

Inaccurate references to the work of others also will cause your readers to be highly suspicious of the reliability of your entire report—and even of your honesty as a writer (see Chapter 11). Inaccurate directions in a set of instructions or procedures might be frustrating at best, disastrous at worst. Considerable problems have resulted when engineers gave measurements in standard units that were assumed to be metric by others. Another kind of inaccuracy might be a claim that is true sometimes but not under all conditions, for example, that water always boils at 100°C. What about purity and variations in atmospheric pressure?

There is also a great difference between fact and opinion. A *fact* is a dependable statement about external reality that can be verified by others. An *opinion* expresses a feeling or impression that may not be readily verifiable by others. The danger comes when opinions are stated as facts:

Fact:	The NR-48 tool features multiple programmable transmitters and a five-station receiver array.
Opinion:	The NR-48 is by far the best piece of equipment for our purposes.

The second statement might be correct but is still only an opinion until supported by verifiable facts. To be strictly honest, the writer should identify it as an opinion unless evidence is presented to support it as fact. In short, make sure that (1) your facts are correct when you write them down and (2) your opinions are presented as such until adequate evidence is provided to verify them.

PRESENT YOUR MATERIAL LOGICALLY

Not only should it be easy to access your document's essential message, but the parts of your information should be sequenced appropriately. Organize your material so that each point or section uses an appropriate overall pattern. If you must follow document specifications, follow those specs, but even then, you may have to present material the way you feel is most effective.

Patterns of Organization

Chronological organization: If readers want to know what progress you have made on a project, what you did on a trip, or how to carry out a procedure, present your material in *chronological* order.

Descriptive organization: If they expect description of equipment or facilities, provide them with a description that moves from *one physical point to another*.

Organization by order of importance: On the other hand, if you have a number of points to make, such as five ways to reduce costs or six reasons to cancel a project, present those points from the *most to the least important*, or vice versa.

Organization by level of difficulty: Perhaps your material needs to be presented in order of *familiarity* or *difficulty*, as when you are writing a tutorial or textbook.

General-to-specific organization: Moving from the *general to the specific* is useful, as in a memo that starts with stringent safety regulations and then provides examples of current unsafe practices.

EXPLAIN THE TECHNICAL TO NONSPECIALISTS

Explaining technical matters to nonspecialists is important to you in at least two ways: (1) your instructors must be satisfied that you thoroughly understand a topic, even though they themselves understand it better than you do; (2) as a professional engineer, you need to be able to explain technical matters clearly so that you can gain people's confidence (and business), their approval for your work projects.

Explaining the technical is a lot about knowing some tools and using them intelligently:

Definitions: One of your most important tools is defining potentially unfamiliar terms. Here is an example of a formal sentence definition: *Noise is a term in both analog and digital electronics that refers to random unwanted perturbation to a wanted signal.*

Examples: For nonspecialists, examples are a big help in understanding the technical: *For example, the "snow" you see on a television or video image is a form of noise.*

Importance: For some readers, discussing the importance of a topic helps them to wake them up and pay attention. For example: *It is important to understand the health effects of noise: Noise pollution can cause annoyance and aggression, hypertension, high stress levels, tinnitus, hearing loss, sleep disturbances, and other harmful effects.*

Uses and applications: People also wake up and pay attention when you explain how something can be used to good advantage. For example: *Dithering is the process of intentionally applying noise to create the illusion of color depth and natural-looking variation in computer-generated images.*

Causes and effects (or reasons); problems and solutions; questions and answers: It helps to discuss these pairs of information types. Here are some examples:

High levels of noise can block and distort the meaning of a message in communications.

Thermal noise can be reduced by reducing the temperature of the circuit.

This is why radio telescopes, which search for very low levels of signal from space, use front-end low-noise amplifier circuits cooled with liquid nitrogen.

Categories: Discussing the categories of a topic can help readers gain a broader understanding of the topic. For example: *The French movie director René Clair points to two types of noise, film dialogue and film noise, and suggests that the latter as well as the former can have meaning and be interpreted.*

In-other-words explanations: When you've written something technical and you are not sure readers will understand, try restating it in simpler words. For example: *European robins living in urban environments are more likely to sing at night in places with high levels of noise pollution during the day. After all, that's when they can hear each other.*

Description: Readers can also gain better understanding simply by being able to visualize some aspect of a topic. For example: *At the north of the sound tube, a massive sculptural work was placed consisting of a giant yellow beam hanging diagonally across the road and a row of smaller red beams alongside the road.*

Process explanation: Going carefully, methodically, step by step through an essential process can also help readers understand. For example:

Whenever a message is sent, someone is sending it and someone else is trying to receive it. In communication theory, the sender is the <u>encoder</u>, *and the receiver is the* <u>decoder</u>. *The message, or* <u>signal</u>, *is sent through a channel, usually speech, writing, or some other conventional set of signs. Anything that prevents the signal from flowing clearly through the channel from the encoder to the decoder is* <u>noise</u>.

History: For some readers, it helps to know the historical background associated with a topic. For example: *After the passage of the U.S. Noise Control Act of 1972, the program was abandoned at the federal level, under President Ronald Reagan, in 1981.*

> **Jellyfish power**
>
> Researchers at Virginia Tech have designed a silicone underwater robot that moves like a jellyfish and that is powered by the water around it enabling it to operate apparently forever.
>
> For details, see the Preface for the URL.

Note   Source for some the examples is Wikipedia's articles on noise and noise pollution (accessed 2012).

MAKE YOUR IDEAS ACCESSIBLE

The organization and structure of a document—specifically, how well the material is laid out in visually accessible "chunks" for the reader—can be readily apparent with the use of (1) hierarchical headings, (2) bulleted and numbered lists, and (3) paragraph length.

HIERARCHICAL HEADINGS

Hierarchical headings enable us to look at the pages of a document and get a sense of its organization and contents. "Hierarchical" indicates that these headings have levels; some are subordinate to others as in a traditional outline. Headings and subheadings act as signposts that help readers get through a document without getting lost. Informative headings also give your document good "browsability"; that is, they help readers quickly find sections that interest them most.

Although practice varies, a common format for the first three levels of headings is as follows:

FIRST-LEVEL HEADING

Write first-level headings in capital letters and put them flush with the left margin. Use boldface to make the heading stand out and separate it from the written material above and below it by at least one line space. (Some formats center the first-level heading.)

Second-Level Heading

Place second-level headings flush with the left margin with at least one line space separating them from text. Capitalize only the first letter of each main word, and make these headings boldface.

Third-level headings. Place third-level headings on the same line as the text they precede. Capitalize them as a sentence would be; use boldface or italics.

Notice that all three levels of headings use a sans serif font (Arial in this case) while body text uses a serif font (Times New Roman in this case).

Numbered Headings.   Some companies and suppliers require numbered or decimal headings. A number system makes it easier to cross-reference other parts of a long report.

FIRST-LEVEL	**1. 0 QUALITY ASSURANCE PROVISIONS**
Second-level	**1.1 Contractor's Responsibility**
Third-level	**1.1.1 Component and material inspection**
Fourth-level	**1.1.1.1 Laminated material certification**

Guidelines for Headings.   When you use headings, watch out for these issues:

1. Phrase headings so that they indicate the topic of the section they introduce. "Background" says nothing; "Background on Noise Legislation" does.

2. As for frequency of headings, try for two to three headings per page, as this book demonstrates.

3. Subordinate headings. If you have these headings, "Types of Noise," "Acoustic Noise," and "Visual Noise," the second and third would be *subordinate* to the first—at a lower level—just as in an outline.

4. Make sure that headings *within the same section* and *at the same level* are parallel in phrasing.

5. Avoid "lone headings," headings that are all by themselves within a section, similar to an outline in which there is an A but no B or a 1 but no 2.

6. Avoid "stacked headings," two or more consecutive headings with no intervening text. Use definitions, transitions, or subtopic overviews to get useful information at these points.

7. When you use third-level headings (the ones that "run in" to the paragraph), start the regular paragraph text as a completely new sentence. Don't "run in" the heading grammatically with the regular paragraph text.

8. Avoid referring to the text of a heading with "this" or other pronouns. That causes readers to pause and wonder "this what?" Apparently, we read headings differently from the way we read regular paragraphs.

To see headings in action, browse the examples in Chapters 5 and 6.

Note   See the chapter in the website companion corresponding to this one for steps on creating heading "styles" for your reports. Styles make formatting your report much easier. For the web address, see the Preface.

USE BULLETED AND NUMBERED LISTS

A well-organized vertical list is sometimes the most efficient way to communicate information: steps in a procedure, materials to be purchased, items to be emphasized, reasons for a decision. In some cases, readers retrieve information from lists more easily than from regular paragraphs. Consider the following:

Problem:

First of all, set the dual power supply to +12 V and −12 V. Next, set up the op-amp as shown in Figure 1. Use a 1 Vpp sine wave at 1 kHz and then plot the output waveform on the HP digital scope. Then obtain a Bode plot for the gain from 200 Hz to 20 KHz.

Revision:

1. Set the dual power supply to +12 V and −12 V.
2. Set up the op-amp as shown in Figure 1.
3. Use a 1Vpp sine wave at 1 kHz.
4. Plot the output waveform on the HP digital scope.
5. Obtain a Bode plot for the gain from 200 Hz to 20 kHz.

1. Use a numbered list for items in a required order. Numbered lists can also be used to indicate an order of importance in your data, such as a list of priorities or needed equipment.

To set up the computer:

1. Connect the monitor to the computer through the monitor port.
2. Connect the keyboard and mouse to the computer through the ASF port.
3. Connect the power supply to the computer.
4. Connect the printer to the printer port.
5. Connect the modem to the modem port.

2. Use checklists to indicate that all the items in your list must be tended to, usually in the order presented:

To set up the computer:

☐ Connect the monitor to the computer through the monitor port.
☐ Connect the keyboard and mouse to the computer through the ASF port.
☐ Connect the power supply to the computer.
☐ Connect the printer to the printer port.
☐ Connect the modem to the modem port.

3. When any list gets longer than ten items, try to break them down into smaller, more manageable sections and give each section its own subheading or lead-in.

4. Use a *bulleted* list when the items are in no required order:

Some of the main concerns of environmental engineering are:

- Air pollution control
- Public water supply
- Wastewater
- Solid waste disposal
- Industrial hygiene
- Hazardous wastes

---

5. Avoid lengthy bulleted lists—over seven items. When list items themselves are lengthy, get some vertical space between them.

6. If you have sublist items in a required order, use lowercase letters. If the sublist items are in no required order, use a less prominent symbol (for example, a clear disc or an en dash).

7. Make sure all lists begin with a lead-in, as the examples here illustrate. Otherwise, readers won't know what to make of your lists.

8. Don't punctuate a lead-in with a colon if the lead-in ends with a verb: for example, *The five priorities we established are*. Punctuate a lead-in with a colon if it is grammatically complete: for example, *We have established the following five priorities:*.

9. Punctuate list items with a period if they are complete sentences or contain embedded dependent clauses.

10. Be consistent with how you capitalize list items that are not complete sentences.

11. Make list items *parallel in phrasing*. For example, if some entries begin with a verb, all entries should do so; if all begin with a noun, all should. This makes for smoother reading and logical neatness. Note how the following list is bumpy due to problems with parallelism:

Lack of parallel phrasing:

Last week we accomplished the following for WW3-a:

- Completed BIU, ICACHE, and ABUS logic design.
- All instruction buffer blocks have had final simulations.
- Written and debugged 75 percent of test patterns.
- Scheduling of first silicon reticules for WW4-a with Vern Whittington in Fab 16.

Revision:

Last week we accomplished the following for WW3-a:

- Completed BIU, ICACHE, and ABUS logic design.
- Ran final simulations on all instruction buffer blocks.
- Wrote and debugged 75 percent of test patterns.
- Scheduled first silicon reticules for WW4-a with Vern Whittington in Fab 16.

Note See the chapter in the website companion corresponding to this one for steps on creating and formatting the different kinds of lists for your reports. For the web address, see the Preface.

CONTROL PARAGRAPH LENGTH

No one, especially in technical fields, wants to read a solid page of wall-to-wall text of difficult material. A busy manager, for example, will want to scan your information as quickly as possible.

When your readers are trying to follow demanding technical information, they are already challenged, and presenting it to them in solid page-long chunks is discouraging. Furthermore, if they need to refer to a point you made or data you presented, they will have trouble finding it.

When editing your work, look for any overly long paragraphs and try splitting them—for example, at a change of topic or type of writing (from concepts to examples). When you split paragraphs, remember that you may have to add a transitional word or phrase.

USE EFFICIENT WORDING

Opinions vary on how much it costs a company for an employee to produce one written page of technical information, but as Chapter 1 states, it can be anywhere up to and beyond $200. With so much writing in industry, costs mount up. A little training in being concise and in sharpening your writing and editing skills can save time and money—plus much time-consuming work for your readers.

INFLATED WORDS AND WORDINESS

Choose the simplest and plainest word whenever you can. Write to communicate rather than to impress, or as the saying goes, "Never utilize *utilize* when you can use *use*."

A few of the more ostentatious—oops, make that showy—words found in engineering writing are listed here, with some plain and efficient counterparts:

Showy	Straightforward	Showy	Straightforward
commence	start	fabricate	make
compel	force	finalize	end
comprises	is	initiate	begin
employ	use	optimal	best
endeavor	try	prioritize	rank
proceed	go	rendezvous	meet
procure	get	terminate	end

Wordiness can also result from using far more words than you need to express an idea. Unkind editors sometimes refer to this as verbiage (by analogy to garbage?). Few of us appreciate hearing a lengthy excuse when a simple "I don't know" is enough. Similarly, your reader is unlikely to thank you for having to plow through when you could have simply said you recommend buying a new computer.

Wordy: I regret to say that at this point in time I basically do not have access to that specific information. . . .

Concise: I do not know. . . .

Wordy: It is our considered recommendation that a new computer should be purchased. . . .

Concise: We recommend the purchase of a new computer. . . .

You can eliminate a lot of wordiness in your writing by training yourself to edit carefully and to make every word count. Look at the following three pairs; you will see which sentences are more efficient and noise-free.

Wordy: It is essential that the lens be cleaned at frequent intervals on a regular basis as is delineated in Ops Procedure 132-c.

Concise: Clean the lens frequently and regularly (see Ops Procedure 132-c).

Wordy: The location of the experimental robotics laboratory is in room 212A.

Noise-free: The experimental robotics lab is in 212A.

Wordy: There are several EC countries that are now trying to upgrade the communication skills of their engineers.

Concise: Several EC countries are trying to upgrade the communication skills of their engineers.

You can also reduce wordiness by avoiding certain pretentious phrases that have unfortunately become common. A good stylebook will give numerous examples, but here are a few that crop up frequently in engineering writing:

Wordy	Efficient
a large number of	many
at this point in time	now
come in contact with	contact
in the event of	if
in some cases	sometimes
in the field of	in
in the majority of instances	usually
in the neighborhood of	about
in view of the fact that	because
in view of the foregoing	therefore
subsequent to	after
the reason why is that	because
within the realm of possibility	possible
in close proximity to	near
in the event that	if
in my own personal opinion	I believe
due to the fact that	because
in close proximity to	near
at that point in time	then
with reference to the fact that	concerning

Check your writing for such unnecessary phrases—as we do in the next sentence. You may ~~often~~ find ~~that there are a number of~~ words ~~contained in your writing~~ that can be ~~safely~~ eliminated without any ~~kind of~~ danger to your meaning ~~whatsoever.~~

REDUNDANCY

Redundancy refers to the use of words that say the same thing, like *basic fundamentals*, or phrases that duplicate what has already been said, as in *They decided to reconstruct a hypothetical test situation that does not exist.* In fact, if you master the art of redundancy, you can make everything you write almost twice as long as need be. Edit your writing once simply looking for redundancy and wasted words—you may be able to reduce word count by as much as 20%.

> *Wordiness*: With reference to the fact that the company is deficient in manufacturing and production space, the contract may in all probability be awarded to some other enterprise.
>
> *Revision*: The company may not be awarded the contract because it lacks production facilities.

What does the preceding revision leave out (other than unnecessary words)? Nothing. Here is a matrix of such possibilities:

Categories of Redundancy

Category	*Wordy version*	*Simpler version*
Redundant adverbs, verbs, and other words	completely finish, completely eliminate, tentatively suggest, connected together, prove conclusively, collaborate together, diametrically opposite, permeate throughout, serves the function of being, has the ability to, come into contact with, exhibits the ability to	finish, eliminate, suggest, connected, prove, collaborate, opposite, permeate, is, can, contact, can
Redundant adverbs and adjectives	totally unique, completely finished, thoroughly complete, bothersomely annoying, productively useful, exactly identical	complete, annoying, useful, identical
Redundant adjectives	complete and total failure, a slender, narrow margin, rectangular in shape	complete failure, narrow margin, rectangular
Redundant adverbs	completely and totally fail, carefully and methodically working, just exactly	fail, carefully working, exactly
Redundant adjectives and nouns	transportation vehicle, tactful diplomacy, successful victory, twenty-four-hour day, time schedule, alternative choices, component part vehicle, alternative choices	vehicle, diplomacy, victory, day, schedule, choices, part, choices

UNNECESSARY PASSIVE VOICE

In the technical world, you must use the passive voice; but when it is misused, it leads to unclear, wordy, and even dangerous writing.

> *Passive-voice problem*: In order to estimate company sales, industry esti-
> mates <u>should first be looked at</u>, because the sales of an individual company
> <u>are often reflected by them</u>.
>
> *Revision*: To estimate company sales, look at industry estimates because
> individual company sales often reflect them.

English has two distinct "voices." The active voice directly states that someone does something: *The engineer wrote the report*. The passive voice turns it around: *The report was written by the engineer*. Thus the active voice emphasizes the performer of the action—the engineer, in our example—while the passive emphasizes the recipient of the action, the report.

Many engineering and scientific writers are told to use the passive voice, that is, to leave themselves out of their writing. They might write *It was ascertained that*... rather than *We found that*..., or *The deadline was met* rather than *We met the deadline*. Management would rather tell you *It has been decided to terminate your employment* than *We have decided to fire you*. (Perhaps such hedging is necessary at times since it helps conceal responsibility and gives us no one to blame!)

The passive voice is certainly appropriate when writing up your research or describing a process, for example. In plenty of instances, you don't want the "doer" to get in the way of your discussion. Also, it's logical to use the passive if the doer of an action is unknown or unimportant:

> *Good uses of the passive voice*:
>
> Electricity was discovered thousands of years ago.
>
> The bridge was torn down in 1992.
>
> The contaminated material is then taken to a safe environment.

Sometimes the passive will give variety to your writing, even if your inclination is to write predominantly in the active voice. In this next example, the passive constructions not only create variety, but they also create better flow by focusing on the *claim*, its *study*, and the subsequent findings and predictions:

> *Effective use of the passive voice for variety and continuity*:
>
> Computer experts claim that general-purpose processors have unpredictable
> execution times due to their use of complex architectural features. This claim
> <u>has now been tested</u> by our group and we have found that the architecture
> really induces little or no unpredictability. Moreover, data gained from our study
> show how the execution times <u>can be predicted</u>. <u>It was also found</u> that ...

In spite of the passive's usefulness, however, the active voice tends to be more efficient. Look at the following pairs, comparing the first sentence to the second:

Passive-voice problem: Control of the flow <u>is provided by</u> a DJ-12 valve.

Revision: A DJ-12 valve controls the flow.

Passive-voice problem: A system for delineating these factors <u>is shown</u> in Figure 5.

Revision: Figure 5 shows a system for delineating these factors.

Passive-voice problem: By switching off the motor when it started to vibrate and looking at the tachometer, the resonant frequency <u>was determined</u>.

Revision: We determined the resonant frequency by switching off the motor when it started to vibrate and looking at the tachometer.

The passive can become especially problematic in procedures or instructions:

Passive-voice problem: The button is pressed twice.

Revision: Press the button twice.

Passive-voice problem: Previously entered data in the database is eliminated by the Edit menu being opened and Select All being chosen.

Revision: Eliminate previously entered data in the database by opening the Edit menu and choosing Select All.

Nowadays engineering writers are getting away from the rigid use of the passive. Sentences become more vigorous, direct, and efficient in the active form. By showing that a *person* is involved in the work, you are doing no more than admitting reality. Also, the active voice gives credit where credit is due. If we read in a progress report that *several references were checked out from the library and 25 pages of notes were taken*, are we as impressed by the energy expended as when we read *I checked out several books from the library and took 25 pages of notes*?

Pedestrian power

A UK company called Pavegen has developed tiles that harvest kinetic energy from pedestrians walking on them.

For details, see the Preface for the URL.

The best policy is to use the active voice when it is the most natural and efficient way to express yourself and when there is no company policy against it. However, don't hesitate to use the passive if the circumstances seem to call for it.

CAMOUFLAGED SUBJECTS AND VERBS

Wordy, awkward, unclear writing occurs when the true subject—or actor—of the sentence is not expressed in the grammatical subject:

Camouflaged subject: Complaints by *employees* about the food served in the company cafeteria have been frequent.

Revision: <u>Employees</u> have frequently complained about the food served in the company cafeteria.

Camouflaged subject: The invention of writing toward the end of the fourth millennium BCE is credited to the *Sumerians*.

Revision: <u>Sumerians</u> are credited with the invention of writing toward the end of the fourth millennium BCE.

Camouflaged subject: An *analysis* of the data will be made when all the results are in.

Revision: We will <u>analyze</u> the data when all the results are in.

The same problem can occur when the main action of the sentence is not expressed in the grammatical verb:

Camouflaged verb: Employees have frequently made *complaints* about the food served in the company cafeteria.

Revision: Employees have frequently <u>complained</u> about the food served in the company cafeteria.

Camouflaged verb: The Sumerians, who lived in southern Mesopotamia (now roughly the lower half of Iraq), achieved the first *creation* of word writing about 3100 BCE.

Revision: The Sumerians, who lived in southern Mesopotamia (now roughly the lower half of Iraq), <u>created</u> word writing about 3100 BCE.

Camouflaged verb: An *investigation* of all possible sources of noise was undertaken.

Revision: All possible noise sources were <u>investigated</u>.

Camouflaged verb: *Acknowledgment* of all incoming messages is performed by the protocol handler.

Revision: The protocol handler <u>acknowledges</u> all incoming messages.

UNNECESSARY EXPLETIVES

Expletives use some form of "it is" or "there is." While they are useful sometimes for emphasis, they too can inflate writing, making it less direct and understandable. Notice the problem version has three expletives!

Weak expletive: <u>It is</u> the results of studies of the central region of the M87 galaxy that have shown that <u>there are</u> stars near the center that move around as though <u>there were</u> some huge mass at the center attracting them.

Revision: Results of studies of the central region of the M87 galaxy show that stars near the center move around as though some huge mass at the center were attracting them.

MIND-NUMBING NOUN STACKS

Another problem, particularly in the technical world, involves jamming three or more nouns together into a phrase, which is called a *noun stack*.

Noun stack: <u>Cocombustion-chamber exit gas temperatures</u> are approximately 2400°F.

Revision: The temperature of gas exiting the cocombustion chamber is about 2400°F.

Noun stack: Install a <u>hazardous materials dispersion monitor system</u>.

Revision: Install a system for monitoring the dispersal of hazardous materials.

Notice how these stacks of five nouns are taken apart and redistributed into phrases. True, the revisions have more words but are more understandable.

WEIRD COMBINATIONS OF SUBJECTS AND VERBS

When you are struggling to express complex technical ideas, it's easy to combine subjects and verbs in strange ways, especially when lots of words come between them. This problem is known as a faulty predication. In this example, its *causes* can't be *devastating—disappearance* can.

Faulty predication: The *causes* of the disappearance of the early electric automobiles *were devastating* to the future of energy conservation in the United States.

Revision: The disappearance of the early electric automobiles was devastating to the future of energy conservation in the United States.

FORMAT YOUR PAGES CAREFULLY

In addition to how you divide information up and how long you make your paragraphs, other factors can also have an impact on your readers. People prefer print that is visually accessible and pleasing. You can easily prevent formatting noise by keeping the following pointers in mind.

MARGINS

Here are some specs for the margins of you document:

- **Margins:** Standard margins are 1 inch all around your page. To get text to fit properly, you can go a little above or below this measurement.
- **Justification:** Avoid full justification, in which the right edge of text is also justified. Instead, use a "ragged" right-hand margin, which makes for easier reading.
- **Binding:** If your report is important enough to be bound like a book, use a wider-than-usual left margin to accommodate the binding and to ensure that the right edge of your text is readable.

TYPOGRAPHY

Typeface is the style of individual letters and characters. *Serif* fonts have small strokes or stems on the edges of each letter, which help guide the eye from letter to letter. Sans serif fonts do not.

Here are some specs for the style of type to use in your documents:

- **Serif fonts:** Use a serif font (for example, Times New Roman) for body text, which is traditionally used by books, magazines, and newspapers.
- **Sans serif fonts:** Sans serif fonts (for example, Arial) are traditionally used for titles and headings. They are also preferred for online text.
- **Font variation:** Documents traditionally use one font for titles, headings, labels, and captions and another font for body text. You will rarely see a third font in all but the most complex documents.

Sans serif: The electric car prototype has regenerative braking, which recharges the supply while decelerating the vehicle.

Serif: The electric car prototype has regenerative braking, which recharges the supply while decelerating the vehicle.

- **Type size:** Standard type size for body text is 10 to 12 point. Use larger type sizes for titles and headings. However, you will rarely see much variation in font size for body text. Regular paragraphs, bulleted and numbered lists, and notices all use the same type size.
- **Capitalization:** Avoid extended text with all capital letters—known as "shouting"—because it is more difficult to read, as the first example illustrates. Use all-caps words for important labels, however:

Problem:

THE GOVERNMENT PLANS TO ESTABLISH A HIGH-LEVEL ADMINISTRATIVE COUNCIL TO COORDINATE SCIENCE AND TECHNOLOGY.

Revision:

The government plans to establish a high-level administrative council to coordinate science and technology.

Good use of capital letters:

DANGER: A 7000V potential exists across the transformer output terminals.

EXPRESS YOURSELF CLEARLY

Engineering is considered a precise discipline (although in reality, as most engineers will admit, it's not always that way). Therefore, engineering writing should be as precise as possible as well. Don't force your readers to work harder than necessary to grasp what you have written. Your sentences must convey a single meaning with no room for interpretation or misunderstanding. If your readers yearn for suspense, they can read a detective story, and if they enjoy different connotations and levels of meaning, they can read poetry.

AMBIGUITY, VAGUENESS, AND DIRECTNESS

The following sections discuss how the meaning of a sentence can be distorted, blurred, or buried in words.

Ambiguity.   The word *ambiguous* comes from a Latin word meaning to be undecided. Ambiguity primarily results from permitting words like *they* and *it* to point to more than one possible referent in a sentence, or from using short descriptive phrases that could refer to two or more parts of a sentence. Consider these following examples.

Ambiguous:	Before accepting materials from the new subcontractors, we should make sure they meet our requirements. *(What or who — the materials or subcontractors?)*
Clear:	Before we accept them, we should make sure the materials from the new subcontractors meet our requirements.
Ambiguous:	The microprocessor interfaced directly with the 7055 RAM chip. It runs at 5 MHz. *(What does "it" refer to?)*
Clear:	The microprocessor interfaced directly with the 7055 RAM chip. The 7055 runs at 5 MHz.
Ambiguous:	Our records now include all development reports for B-44 engines received from JPL. *(What was received from JPL — the reports or the engines?)*
Clear:	Our records now include all B-44 engine development reports received from JPL.
Ambiguous:	After testing out at the specified high temperatures, the company accepted the new chip. *(Did the company or the chip test out at the high temperatures?)*
Clear:	The company accepted the new chip after it tested out at the specified high temperatures.

Vagueness.   If ambiguity involves more than one meaning, vagueness involves no useful meaning at all. What would you think if your doctor told you to "take a few of these pills every so often"? Vagueness is also caused by abstract words—they are not precise. Try to avoid abstractions like *substantial amount*, *corrective action*, and the annoyingly lazy and unspecific *etc*. Replace them with terms that have exact meaning such as *in three days, $8,436.00, replace the altimeter*. Here are two more examples of vague writing:

Vague:	The Robotics group is several weeks behind schedule.
Useful:	The Robotics group is six weeks behind schedule.
Vague:	The CF553 runs faster than the RG562 but is much more expensive.
Useful:	The CF553 runs 84% faster than the RG562 but costs $2,840 more.

As you can see, vague writing may require fewer words, but at the expense of precision. On the other hand, vagueness can be an asset to people who don't want to reveal too much—or who have nothing to reveal. The following satirical "Progress Report for All Occasions" has been going around industry for some years now; it is a monument to vague writing:

Vague:

During the report period that encompasses the organized phase, considerable progress has been made in certain necessary preliminary work directed toward the establishment of initial activities. Important background information has been carefully explored and the functional structure of components of the cognizant organization has been clarified.

The usual difficulty was encountered in the selection and procurement of optimum materials, available data, experimental data, and statistical analysis, but these problems are being attacked vigorously, and we expect that the development phase will continue to proceed at a satisfactory rate.

Directness. Being as direct as possible in your writing lets your reader grasp your point quickly. A busy technical reader wants access to your information quickly and easily. The most important part of your message should come at the beginning of your sentences and paragraphs—unless for the sake of coherence you need to use the old-to-new pattern (discussed in the next section):

Indirect:	After a long and difficult development cycle due to factory renovation, the infrared controller will be ready for production in the very near future.
Direct:	The infrared controller will be ready for production March 4. Its development cycle was slowed by the factory renovation.
Indirect:	Fred has been busily working on this project. This past week he also reworked the logic diagrams, rewired the controller arm, and redesigned all of the RIST circuitry.
Direct:	Fred redesigned the RIST circuitry on Thursday. He also reworked the logic diagrams and rewired the controller arm last week.

COHERENCE AND PARAGRAPH STRUCTURE

When you review your rough draft, look for ways to strengthen the organization and flow of your ideas. Do this kind of review at the level of a whole paragraph and a whole group of paragraphs:

- Strengthen transitions between major blocks of thought, such as between paragraphs or groups of paragraphs. (See Chapter 2 for more on transitions.)
- Experiment with the old-to-new pattern: Begin a sentence with the "old" topic of the preceding sentence and put new information afterwards in that same sentence. Repeat—don't vary—word choice for key terms in the discussion.
- Add topic sentences (particularly the overview kind) to paragraphs where appropriate.
- Check the logic and sequence of paragraphs or groups of paragraphs. To do so, mentally label each paragraph or paragraph group with one or two identifying words. This method enables you to get the "global picture" more easily.
- Break paragraphs that go on too long and challenge the reader's attention span.
- Consolidate short paragraphs that focus on essentially the same topic. Too many paragraph breaks can have a fragmented and distracting effect.
- Interject short overview paragraphs at the beginning of sections and subsections.

Using these strategies guides readers through your report, showing them what lies ahead, what they have read previously, and how everything fits together.

The root of the word *coherence* is *cohere*, meaning to stick together, and a cohesive group of sentences does just that.

In a coherent paragraph, all the sentences clearly belong where they are because they address only the topic of the paragraph and are logically connected to one another. In a complete report, coherence means how well the report takes the reader through its pages by means of transitional devices and how everything focuses on the subject of the report.

Transitional techniques help connect ideas, distinguish conditions or exceptions, or point out new directions of thought. Simple words like *therefore, thus, similarly*, and *unfortunately* eliminate ambiguity by helping a reader interpret your information. Neglecting these techniques means creating noise. Consider the problem sentence below. Both sentences in the problem version are grammatically correct and contain important facts, but how are these facts related? Now notice how the three attempted revisions guess at that relationship which the problem version does not indicate:

Problem: The group's long-range plans for the S-34B project have been extended. The completion date for the project is as originally planned.

Revision possibilities:

The group's long-range plans for the S-34B project have been extended. Nevertheless, the completion date for the project is as originally planned.

The group's long-range plans for the S-34B project have been extended. <u>Unfortunately</u>, the completion date for the project is as originally planned.

<u>Even though</u> the group's long-range plans for the S-34B project have been extended, the completion date for the project is as originally planned.

You can prevent readers from having to guess at connections and achieve coherence by using several techniques:

- Make sure each sentence clearly relates to the one before it and after it. If none of the words in a sentence indicates any connection to the sentences before and after it, you've got a problem!
- If each sentence makes some statement about the same topic, find a way to start most of the sentences with that topic toward the beginning of those sentences. (Don't vary word choice for key words.)
- Use the old-to-new technique when possible: the new subtopic often occurs in the second half of a sentence; echo that subtopic at the beginning of the next sentence. (Again, don't vary word choice for key words.)

    These last two points may seem contradictory. That's because the flow of topics (topic strings) can be *continuous* as in the first example or *shifting* as in the second example.

- Use transition words and phrases to further strengthen the connection between sentences.

In the revised version of this example, a continuous topic string is used; each sentence makes a statement about the 125-H CRT:

Poor Coherence

A significant disadvantage of the 125-H CRT is its high power consumption. To produce the high voltages and currents that are necessary to drive and deflect the electron beam, the tube requires substantial power. The 125-H CRT is inefficient because only about 10% to 20% of the power is converted into visible light at the surface of the screen. Portable display devices that run on batteries, where lower power consumption is necessary, are not suitable for the 125-H. We should consider other options before committing to purchase the 125-H.

Effective Coherence with a Continuous Topic String

A significant disadvantage of the 125-H CRT is its high power consumption. *This* tube requires substantial power to produce the high voltages and currents that are necessary to drive and deflect the electron beam. *In addition*, the 125-H is inefficient—only about 10% to 20% of the power used by the tube is converted into visible light at the surface of the screen. *Thus*, the 125-H is poorly suited for portable display devices that run on batteries, where lower power consumption is necessary. *Because of this drawback*, we should consider other options before committing to purchase the 125-H.

In the original version, notice how the second and following sentences begin with new words, causing us to have to wait to see the connection to the previous sentence. In the revision, notice that in sentence 2 the word "This" makes a strong connection to the preceding sentence. In sentences 3 and 4, "In addition" and "Thus" are transitional phrases that help the flow. In the last sentence, "Because of this drawback" is a powerful transition that summarizes the idea of most of the preceding sentences and connects it as the reason for considering other options.

Here is an example in which the flow is predominantly shifting:

Effective Coherence with a Shifting Topic String

The most important part of a solar heating system is the <u>solar collector</u> whose main function is to heat water to be used in space heating. There are various <u>types</u> of collectors. However, the <u>flat-plate collector</u> is the most common and the focus of the following discussion. A <u>flat-plate collector</u> consists of a box-shaped black plate absorber covered by one or more transparent layers of glass or plastic with the sides and the bottom of the box insulated. <u>These layers of glass</u> or plastic have an intervening air space that produces the heat-trapping effect. Water is <u>heated</u> as it circulates through or below the absorber component, which is heated by solar radiation.

Notice how this paragraph moves from general to specific with each sentence. We start at solar collectors, move to a specific type, then to its construction, and finally the layers of glass or plastic and their heat-trapping effect.

MANAGE YOUR TIME EFFICIENTLY

Few engineers feel they have enough time to do the writing required of them. Often a memo is hastily churned out, or a report is rapidly thrown together and tacked on the tail end of a project. As with anything done in a hurry, the results are usually not the best. As the pressure to get a piece of writing out increases, sloppiness—that is, noise—also increases. Rather than leaving your writing to the last minute, consider it just as much a part of your professional activities as designing, building, and testing.

FINDING AND USING TIME

You can find time to spend on careful writing and editing in a number of ways, but most are not too attractive. Get to work earlier, or take work home (plenty of successful engineers do). Use breaks to concentrate on your writing tasks. Designate a specific time each day for your writing. Write on your laptop computer at airports, in flight, on trains, in hotels, or in waiting rooms.

However, it's much more practical to make your written work an organic part of your day. Assign brief chunks of time for short memos and letters or for small sections of a report. Designate larger chunks of time to concentrate on longer writing tasks.

OUTLINES, DEADLINES, AND TIMELINES

When you have to write anything over two pages long, it's useful to first spend some time making a rough outline. This outline does not have to be final, but it will help you divide your task into smaller sections that can then be written separately at different times, and not necessarily in any order.

Even if you do not have a deadline for completing a document, establish one. Estimate how long you expect the job to take and schedule back from there. You might even draft a timeline for yourself, showing each date by which you should have completed specific parts of the paper. (See Figure 3-2.) Always allow yourself enough time at the end to review and edit the entire document.

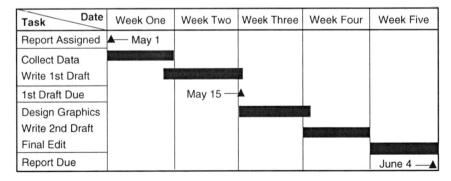

Figure 3-2   The timeline you make for your writing project can be as simple or as detailed as you wish. Make sure you have all your important tasks and due dates down, however, and then do everything you can to keep to them.

EDIT AT DIFFERENT LEVELS

Rather than expecting to randomly find anything in need of improvement, many writers prefer a more methodical approach to editing. First, check your document for *technical accuracy*. Then decide what "writing levels" to approach your editing on, and go through your document at least once on each level.

LEVEL 1

In your first pass through your rough draft, check the overall content, organization, coherence, readability, and accuracy of content. Are headings, subheadings, lists, and graphics used effectively and consistently? Are the parts of the document arranged correctly? Fix these high-level things before working on things like punctuation errors—why fix errors in sentences that will be cut?

LEVEL 2

In your second pass, check for paragraph and sentence length and structure, wordiness, and word choice. Is the tone of your document appropriate? Have you used the active voice where possible? How about transitions, parallelism, and emphasis where called for? These are mid-level considerations, which also should come before fixing things like punctuation errors. No sense fixing what may be cut or overhauled.

LEVEL 3

In your third pass, you check for the nitty-gritty problems like mechanics, spelling, punctuation, typos. Again, a good word processing program will provide you with suggestions on spelling and grammar; however, *you* must make the final choices on many of these options. You might also call upon the services of a friend or colleague who is well grounded in these basics.

LEVEL 4

When you have worked through the previous three levels of review, made the necessary changes, and now have a near-final document, consider its appearance. Are specifications (if any) followed? Is it the right length? Have you used the best font size, margins, and spacing? Are headings, subheadings, lists, and graphics used effectively and consistently? Is the title page attractive? How about the "packaging" of the document, such as paper, binding, and covers?

> **Breathing power**
>
> Engineers at University of Wisconsin-Madison are developing a device that uses low speed airflow like that caused by normal human respiration to generate electricity.
>
> For details, see the Preface for the URL.

SHARE THE LOAD: WRITE AS A TEAM

Not many engineers write lengthy reports by themselves. Technical people work together as teams for research, design, development, and testing. They often find they must team up to write proposals, manuals, and completion reports. Team writing is not always easy, especially when people with different degrees of writing ability or ego investment are involved, or when team members are torn between team responsibilities and other duties. If your group plans the team project carefully, however, it can be a great experience since as a team you will be tapping into far more knowledge, skill, and creativity than you can bring to a project alone.

FIVE POSITIVE APPROACHES

When you work on a team project or help put together a long written document with others, be prepared to do the following:

- Communicate
- Coordinate
- Collaborate
- Cooperate
- Compromise

This list might seem obvious, but many teams fail to reach their potential because some members have difficulty in following it. Some people even view *collaboration* and *compromise* negatively rather than positively in the context of team activities.

Communicate. Obviously, very little teamwork is possible without frank and open communication. Members of the team must create an atmosphere that enables free discussion at all times. Channels of communication (email, telephone numbers, mail addresses, and meeting times and places) must all be open.

Coordinate. Since team members are not often physically working together, everyone must know what the others are doing, who is responsible for what, when the next deadline or meeting is, as well as other tactical details. Often one member of a team is appointed as the coordinator, and if that person does the job well, a minimum of frustration, repetition, or uncertainty will occur.

Collaborate. The Latin root of this word means "to willingly labor." In a team setting, it means just that—to willingly assist one another. In the spirit of collaboration, for instance, assist a partner who is overloaded and needs help. Freely share your own work with the other team members and work together at creating a final document that is unified and seamless.

Cooperate. Cooperation is essential to the smooth working of any team project. If the project has a designated leader, do all you can to cooperate with that person and to accept his or her decisions, deadlines, changes, or reassignments. If such executive actions by the team leader cannot be the result of open discussion, cheerfully accept a decision you have no control over.

Compromise. This word has two meanings, and only one of them is somewhat derogatory. The other meaning refers to making mutual concessions in order to reach a goal. In practical teamwork, this means you may sometimes have to give a bit on something because doing so will help the team reach its objective. As much as possible, compromise should be the outcome of open and friendly communication.

PRACTICAL TEAM WRITING

Here are two practical ways your team can produce a successful written document:

- **Everybody writes:** Divide the length of the assignment by the number of people involved and get each to write his or her share. Individuals will do any research needed for their own section and should write and edit it. Then the document can be "glued together."

  This method may not result in an efficient or effective product. Individuals bring their personal writing style, vocabulary, quirks, and weaknesses to their part; their material may overlap with other parts of the report; transitions between sections will likely be absent. Your team will need a project manager to push things along and settle disputes. You will also need a final editor who can take the completed draft and mold it into a coherent and useful document.

  However, the good thing about this strategy is that everybody gets to write.

- **Everybody specializes:** In some ways, the best strategy to produce a team document is to assign tasks according to individual members' strengths and interests:

  - One person acts as project manager, organizing and assigning tasks, checking to ensure the project is on schedule, and even refereeing disputes.

  - Another team member gathers the needed information for the document, writes notes, and puts together a very rudimentary draft.

  - Another member, the "strong" writer, generates a working draft. Ideally, this person is good at writing, enjoys writing—and has read this book.

  - Still another team member with editing skills can act as quality control officer, reading, checking, editing, and in general perfecting the document while working closely with writers.

The unfortunate thing about this strategy is that not everyone gets to write—which might be okay in industry but not in an engineering writing course.

For both strategies, other specializations are likely to be needed: someone who is good at graphics; someone who is well versed in the topic to research difficult technical areas; someone who has a flair for graphic design to assemble the final copy.

EXERCISES

1. Think of some significant communication events you have experienced in the past several months at work or in class. What kinds of audiences were involved? Did a lack of clearly defined audience and purpose cause noise in the communication process?

2. Look inside the back cover of an IEEE or other technical journal for advice for authors who wish to publish in that journal. Are specifications given for such details as abstracts, length, headings, margins, columns, graphics, size of print, references, and so on?

3. Find a government or industry report on a subject that interests you. Who is the assumed audience? How does the report get to the point right away—if it does? How useful are the headings and subheadings? How do divisions and paragraph length add to the accessibility of the information? Could any of the information be better presented in list form? Do you notice any ambiguity, wordiness, unnecessary technical jargon, and nouns that could be turned into verbs?

4. Look back on any solo writing project you have done. How long did it take you? Were you working under a deadline? How much time each day did you spend planning, writing, and editing? How could you have done the project better?

5. Look back on any team writing projects you have been involved in. How were tasks or sections delegated? Were you satisfied with the completed document? Was whoever assigned you the task satisfied with your work? What factors would have enabled you to do an even better job?

BIBLIOGRAPHY

Alred, Gerald J., Charles T. Brusaw, and Walter E. Oliu. *The Technical Writer's Companion.* New York: Bedford/St. Martin's, 2002.

Larson, Kevin. The Technology of Text, *IEEE Spectrum*, May 2007, pp. 26–31.

Microsoft Manual of Style for Technical Publications. New York: Microsoft Corporation, 2004.

Miller, G.A. The Magical Number Seven, Plus or Minus Two: Some Limits on Our Capacity for Processing Information. In *The Psychology of Communication.* New York: Basic Books, 1967.

Nadziejka, David E. The Levels of Editing Are Upside Down. *Proceedings of the International Professional Communication Conference*, pp. 89–93, September 28–October 1, 1984.

Reep, Diana C. *Technical Writing: Principles, Strategies, and Reading.* New York: Pearson Longman, 2009.

4

LETTERS, MEMORANDA, EMAIL, AND OTHER MEDIA FOR ENGINEERS

A recent . . . survey found that the average employee spends nearly an hour a day handling e-mail chores. For managers, e-mail tasks usurp closer to two hours each day. It's no wonder people are complaining about e-mail fatigue.

Paul McFedries, "The Age of High (Tech) Anxiety,"
IEEE Spectrum, June 2003, p. 56.

The finest eloquence is that which gets things done.

David Lloyd George, 1863–1945.

As a professional engineer, you need to get familiar with the style, format, and organization of business communications. (With contemporary down-sized organizations, it may be a while before you rise high enough in the firm that you can rely on a secretary.) This chapter explores strategies for deciding which medium of communication to use and then moves on to discuss format, style, and organization of business letters, memoranda, and email. The chapter concludes with writing-style issues that apply to any of the media described here.

WHICH TO USE?

Working professionals have at their disposal a variety of communication media. If you have a question for someone in your building, run down the hall and ask in person. If it's to someone within your organization but at a different location, send email or write a memo. If it's to someone external to your organization, send an email or text message or write a business letter. If it's urgent or informal, make a telephone call or send an email or text message. If you want to build your reputation or that of your organization and if you want to join into ongoing discussions in the engineering community, use social media such as Facebook, LinkedIn, blogs, forums, and Twitter.

PHONE, TEXT MESSAGE, OR PAPER?

The decision whether to use telephone or face-to-face communication as opposed to written communication is fairly obvious. In telephone or face-to-face communication, these are the issues:

- **Permanent record.** There is no record of what transpires in your phone conversation, unless you record the conversation (which can have a chilling effect). Text messages can be saved, but they are so limited that they are not a good option for professional activity.
- **Availability of the recipient.** Mobile phones have reduced the problem known as "telephone tag." Still, recipients of a communication may not be in their offices or able to answer their mobile phone. Text messages remain available until the recipient can review them, but again their limitations are such that they are not a practical option.
- **Attitude of recipients.** Recipients may not take the phone, text-message, or in-person communication as seriously as they would if it were in writing.
- **Purpose, length, and complexity of the topic.** Some topics are just too much for a conversation. For example, you can't present details of product specifications or a proposal over the phone.

EMAIL, INSTANT MESSAGES, OR PAPER?

The decision gets harder when you choose between the various forms of Internet communication and print. If you use email, you may wonder why you should bother with phone calls, business letters, or memos at all. Email and instant messages eliminate the bother of stamps, envelopes, and mailboxes—not to mention the delay in delivery and response. Unlike telephone communication, email doesn't require its recipients to be in the right place at the right time—they can read it when they are ready. And, unlike telephone communication, email constitutes a record of the communication, although viewed by some as unofficial. However, print remains the preferable media in certain instances, and sometimes the only communication media. Here are the issues to consider:

- **Recipients.** Obviously, if recipients don't have email, can't readily access email, or just do not like it, printed letters or memos are necessary.

- **Need for reply or forwarding.** If the letter or memo contains pages that the recipient must fill out and send, hardcopy may be preferable.

- **Security issues.** As Ed Krol pointed out in *The Whole Internet User's Guide and Catalog* in the early days of the Internet, you can assume that any email you send has a chance of being seen by anyone in the world. Think twice about sending confidential information (new product specifications, confidential data about a project, or sensitive information about a colleague) by email.

- **In-person discussion of the memo.** If the message must be used in a face-to-face situation, print may be preferable. If everybody must print the memo for the meeting, you might as well send it in print and thus eliminate a potential snag.

- **Importance or length of the information.** For some, even now deep into the age of the Internet, email lacks the feeling of settled, established information. It seems light, ephemeral—not a medium for serious business. Some people are less likely to take an electronic message seriously than they are a hardcopy memorandum or letter.

- **In-your-face factor.** For some, a printed memo sitting on their desk just cannot be avoided. Of course, that depends—for some professionals, hardcopy mail is more inconvenient than email. Ultimately, you have to base your decision on which medium your colleagues are most in the habit of using.

Standard advice is not to email when you are mad, when you are drunk, when it's 3 a.m., when you are bored, or when you just feel like gossiping.

Dave Pollard is one of many voices arguing that email is now overused and overabused. (See the links to his web pages at the companion website for this book.) He recommends not using email in the following situations:

- To communicate bad news, complaints, or criticism
- To seek information that is not simple and straightforward
- To attempt communications that will require lots of back-and-forth exchanges
- To seek approval on something that is complicated or controversial
- To send complicated instructions
- To request comments on a long document
- To request information from a group on a recurring basis
- To achieve consensus

- To explore or brainstorm a subject or idea
- To send news, interesting documents, links, policies, directory updates, and other "FYI stuff"

For most of these situations, Pollard recommends a phone call, face-to-face meeting, or audio/video conferencing. For collaboration on long documents, he recommends screen-sharing technology (for example, web-based meeting software applications such as Adobe WebConnect) if face-to-face meetings are not possible. For interesting news and documents, he recommends posting to a wiki or a social-networking site (such as LinkedIn) and using RSS newsfeed capabilities.

On the other hand, collaborative project management tools such as Google Docs, Google Sites, and Microsoft Sharepoint concentrate communications within the framework of a project. Project communications stay within the project rather than coming to your regular email.

LETTER OR MEMO?

Memoranda are written communications that stay within an organization. Business letters are written communications to recipients who are external to the organization of the sender. Of course, some internal communications are in the form of business letters—for example, those letters that the CEO sends out once or twice a year to all employees. Obviously, email has encroached on much of the territory formerly owned by printed letters and memos, but the advice in the preceding section on when not to use email still applies.

> ### Sole power
>
> An electrical engineer at Louisiana Tech has developed a device that harvests power from a small generator embedded in the sole of a shoe. The device converts a piezo-electric charge into usable voltage for charging batteries or for directly powering electronics such as mobile phones.
>
> For details, see the Preface for the URL.

FACEBOOK, LINKEDIN, TWITTER, BLOG, OR FORUM?

Social media provide excellent resources to build your professional presence, promote your organization, and join in the community of engineering discussions. As pointed out by Dave Pollard above, social media provide a much better venue for certain kinds of communication than email which is already way overloaded.

See Chapter 12, "Engineering Your Online Reputation," for details on social media tools.

WRITING STYLE FOR BUSINESS CORRESPONDENCE

Regardless of the medium you use for your business correspondence, most of the guidelines for writing style are the same. Whether you are writing a business letter, memorandum, or email, the following recommendations are equally valid:

- **Indicate the topic in the first sentence.** Don't assume recipients read your subject line (however clear and compelling it may be). State the topic and purpose of your communication in the very first sentence.

- **Identify any situation or preceding correspondence to which your communication responds.** In the first paragraph, establish the context by referring to any previous meeting, phone conversation, or correspondence.

- **Provide an overview of the contents of the communication.** If the letter, memo, or email is lengthy, provide an overview of the contents—nothing more than an informal list in a sentence within the first paragraph.

- **Keep the paragraphs short.** Ideally, paragraphs in business correspondence should not go over five to seven lines. Readers are less willing to wade through long, dense paragraphs in business correspondence than they are, for example, in textbooks or formal reports.

- **Use headings for communications over a page in length.** If your communication is more than a page or two and if the information in it is like that in a report, use headings to mark off the boundaries where new topics start. (See Chapter 3 for more on headings.)

- **Use lists and graphics as you would in a report.** Business correspondence can at times resemble reports; writers use the same sorts of headings, lists, and graphics in their letters and memos. Look for ways to create lists, particularly in long paragraphs. Similarly, use graphics and tables in your correspondence just as in regular reports.

- **Be brief, succinct, to the point.** Brevity is never so important as it is in business correspondence—and still more so in email. Readers lack patience with unnecessary background and wordiness.

- **Use an interactive style in memos and email.** Be as informal as the situation allows. Whenever appropriate, use the "you" style of writing—avoid the impersonal third-person and passive-voice styles.

- **Indicate any action necessary on the part of the recipient.** Let readers know what you expect them to do as a result of reading your correspondence. What actions should they take after reading your letter, memo, or email? Fill out a questionnaire? Where is it located? Where should they send it? Make sure that all details like these are clearly and specifically explained.

COMMUNICATION STRATEGIES FOR TRICKY SITUATIONS

In your business communications, you will at times have to deal with some tricky situations such as the following:

- **Tell the boss "no".** If the boss has directed you to do something stupid, unethical, or even illegal—knowingly or unknowingly—your job is to explain it to that boss, preserve your position in the organization, and most likely protect the organization as a whole.

- **Request reimbursement.** If you have had a bad experience with a faulty product or service, don't just angrily fly into the details from line 1. Write a good introduction as described previously. Segment your discussion into sections objectively describing the details of the problem, explaining what reimbursement you expect and why, and then concluding by stating your hopes that you can continue doing business with the recipient.

- **Give bad news.** Sometimes, you cannot approve what a client or customer has requested. Say no *after* you've given the reasons and then find a way to keep that person's business, possibly by providing some token of good will. Bad news is often "buffered" by some neutral statement so that the reader will read the rationale. But there are good and bad buffers. Here's a questionable one:

  Bank Two is delighted to have you as a mortgage customer and hope to retain you as a valued customer.

  It's a buffer all right, but there is no indication of what's coming. The reader must surely be wondering what's this about. A nonbuffered approach would bluntly state that the reader's company cannot take on any further debt financing. However, in this next example, the buffer is the fact that the writer does not reveal what those conclusions are:

  I am writing to inform you of our conclusions reached regarding your company's pursuit of further debt financing.

- **Admit a mistake.** If you make a mistake, you must find a professional way to admit that mistake, explain why it happened, outline what you'll do to ensure it won't happen again, and protect your position and reputation in the organization. A buffer would be helpful here as well; it would get readers to keep reading rather than throw up their hands in disgust.

- **Assert that you did not make a mistake.** In some organizations when a mistake is made, fingers start pointing (blame-throwers). If you are seen as a culprit, you've got to find a professional way to defend yourself.

- **Issue an unpleasant directive.** If as a supervisor you must issue an unpleasant directive (for example, weekend work to meet a deadline), you must find a way to do that without sounding like a dictator and by showing that it is in everyone's best interest.

Obviously, not all of these situations require written documents. Sometimes, it's best to let a problem go away. However, if some record must be kept, your writing skills and awareness of strategies become very important.

Note See examples of business communications that address these tricky situations in the companion website for this book. See the Preface for the URL.

BUSINESS LETTERS: COMPONENTS AND FORMAT

As suggested earlier, the common business letter (printed on real paper!) is not dead. Face-to-face, telephone, and email communications are just not right for certain kinds of correspondence. Use a hardcopy letter,

- when you want to make sure that the recipient receives it, takes it seriously, studies it at length, and acts appropriately upon it.
- when the communication is long and packed with information.
- when you want a permanent record of the communication.
- when organizations, such as the U.S. Department of Defense and the Department of Health and Human Services, have strict guidelines on sending certain communications as printed hardcopy.

Use the following design suggestions for business letters—professional communications external to your organization.

STANDARD COMPONENTS OF BUSINESS LETTERS

The following describes standard components for business letters, most of which are illustrated in Figure 4-1.

- *Company stationery and logo.* If you use company stationery, begin your letter about an inch below the logo. Don't use logo stationery on following pages; use the matching stationery without the logo. If you are an independent consultant, design your own logo with an interesting type style, maybe some combination of bold and italics, and maybe horizontal lines above or below your name, title, and address. (See the examples throughout this book.)
- *Heading.* The heading contains your (the sender's) address and the date. If you're using letterhead stationery, only the date is needed.

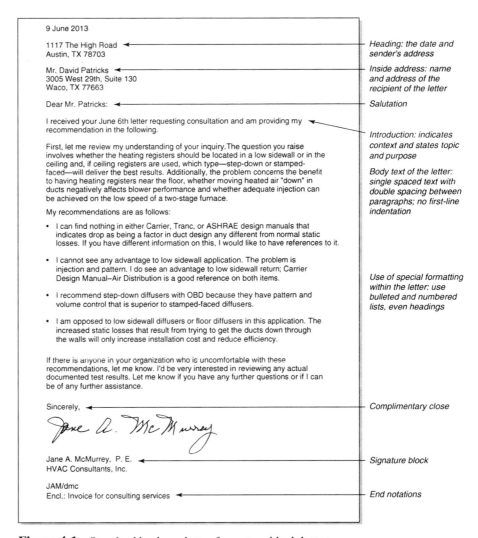

Figure 4-1   Standard business letter formats—block letter.

- *Inside address.* This portion includes the name, title, company, and full address of the recipient of the letter. Make this the same as it appears on the envelope. This element becomes important when secretarial staff discards the envelope.
- *Subject line.* Some business-letter styles make use of a subject line, the same kind that you see in memoranda. This element announces the topic, purpose, or both of the letter—for example, "Request for copyright status on the XI1 documentation" or "In response to your request for copyright status." (See Figure 4-2 for an example.)

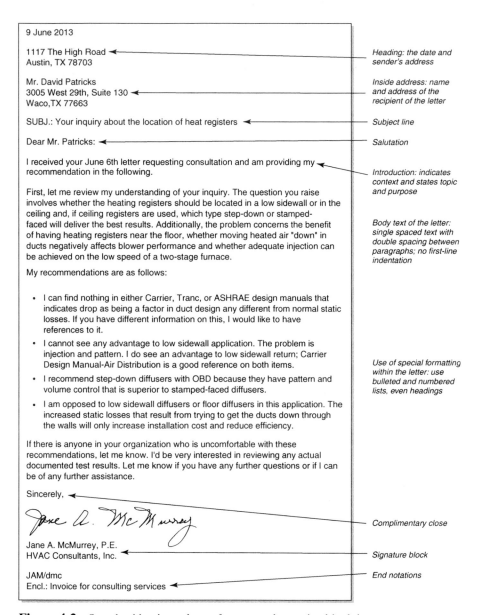

Figure 4-2   Standard business-letter format—alternative block letter.

- *Salutation.* This is the "Dear Sir" element. In contexts where no obvious recipient exists or where the recipient does not matter, omit the salutation. If you must include a salutation but don't have a specific name, call the recipient's organization (also ask for title and department name), or create a department or group name that is reasonably close. For example, use "Dear Recruitment Officers:" If all else fails, use the infamous "To Whom It May Concern:". Notice that the salutation for business letters is punctuated with a colon. (A comma implies a friendly, nonbusiness communication.)

- *Body.* The body begins just after the salutation and continues until the complimentary close. Text is single spaced; first lines of paragraphs are not indented; and double spacing is used between paragraphs. (For writing strategies and style to use in the body, see "Writing Style for Business Correspondence" earlier in this chapter.)

- *Complimentary close.* In letters where there is no interpersonal action, this "Sincerely yours" element can also be omitted. Capitalize only the first word and punctuate with a comma.

- *Signature block.* This is the blank area for the signature, followed by your typed name, title, and organization. Don't forget to include those letters that identify the degree or title that you worked so hard to earn. Below your name, include your title and the name of your organization.

- *End notations.* These elements are the "Cc:" and "Encl:" abbreviations below the signature block. In Figure 4-1, the first set is the initials of the sender; the second set, the typist (for example, "JAM/dmc"). Labels like "Encl.," "Enclosure," or "Attachments" indicate that other documents have been attached to the letter. If you want, you can specify exactly what you've attached: for example, "Encl.: specifications."

    "Cc:" followed by one or more names indicates to whom a copy of the letter is sent. "Bcc:" is an office-politics stratagem that identifies "blind" recipients. If you receive a letter with "Bcc:" at the bottom, the people whose names follow "Bcc:" do not know that you received the letter, nor do they know that you know that they received the letter.

- *Following pages.* If you use letterhead stationery, use the matching pages (the same quality of paper but without the letterhead) on following pages. On following pages in professional correspondence, use a header like one of those shown in Figure 4-3.

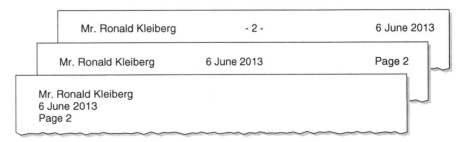

Figure 4-3   Three separate formats for following pages in business letters.

COMMON BUSINESS-LETTER FORMATS

Traditionally, business letters have used one of four standard formats: block, semiblock, alternative block, and, more recently, simplified formats. These formats vary according to which elements are present (for example, a salutation) and where they are placed on the page (for example, on the left or right margin).

- *Block format*—The easiest and most commonly used; all elements are flush left. (This format is shown in Figure 4-2 with the addition of a subject line.)

- *Semiblock format*—Similar to the block format except that the heading, complimentary close, and signature block are at the right margin.

- *Alternative block format*—The same as the block or semiblock format except that it adds a subject line. See Figure 4-2. Use this format because it includes the valuable subject line.

> **Pavement power**
>
> A Spanish tech company has developed a product called iPavement —sidewalk paving stones that double as WiFi hotspots. Power and internet access are supplied to each stone via a hard-wired 1,000-watt cable.
>
> For details, see the Preface for the URL.

- *Simplified format*—The same as the block format except that it omits the salutation.

Note See the corresponding chapter in the website companion for examples of these other letter formats. For the web address, see the Preface.

For communications that involve no interpersonal interaction, the simplified and the alternative formats are best. For serious professional communications, such as proposals or employment letters, stick with the block or alternative block format. The semiblock is rarely used any more and is tedious to format.

BUSINESS MEMORANDA

For communications internal to an organization, use the memorandum format. The actual contents of a memo can be very much like those of a business letter or like those of a short report—the key is the memorandum format.

As with formal printed business letters, you may wonder whether printed memoranda are necessary in the age of the Internet. Your organization may have strict policies about when email is not permissible. The central concern is confidentiality; any email is likely to be seen by anybody. For that reason, a job offer, a request for a raise, and complaints about an employee are good examples when *not* to use email.

STANDARD COMPONENTS OF MEMORANDA

The format for memoranda is much simpler than that for business letters. Figures 4-4 and 4-5 illustrate the standard components.

- *Heading—DATE.* While formats vary, put the date you send the memo in the header. The example in Figure 4-5 shows it as the third line in the heading; in some designs it is the first line, as in Figure 4-4.
- *Heading—TO.* Put the name of the recipient or the group name in this slot. The level of formality is very apparent here. You can put "Elizabeth," "Elizabeth

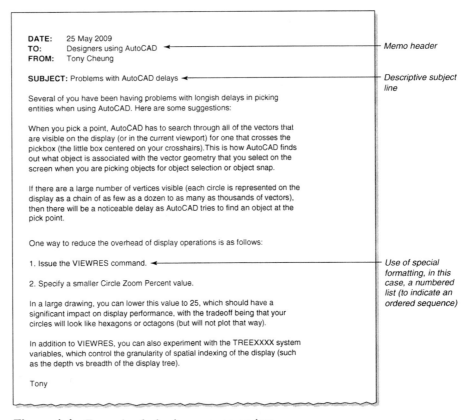

Figure 4-4   Example of a business memorandum.

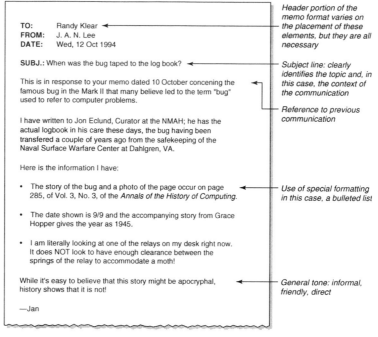

Figure 4-5 Example of a business memorandum.

Bennet," or "Ms. Elizabeth Bennet, Director of Personnel," depending on your familiarity with the recipient and the formality of the situation.

- *Heading—FROM.* Put your own name or the name of the person or group for whom you are writing the memo in this slot. Once again, familiarity and formality dictate whether to put just your first name, your full name, or your full name and title. Traditionally, as the writer of the memo, handwrite your initials or first name just after your printed name.

- *Heading—SUBJECT.* In this slot, place a phrase that captures the topic and purpose of the memo. For a survey of grammar-checking software, the subject might be "Results of our survey on grammar-checking software." The actual label for this element varies: Some styles use "RE:" or "SUBJ.:". If your memo is a response, phrase the subject line accordingly. For example: "Re: your request for a survey of piezoelectric devices."

- *Body.* The body begins just after the heading. Text is single spaced; first lines of paragraphs are not indented; and double spacing is used between paragraphs. (For writing strategies and style to use in the body, see "Writing Style for Business Correspondence" earlier in this chapter.)

- *Signature block.* In formal memoranda styles, writers actually insert the same kind of complimentary close and signature block that you see in business letters. Otherwise, the signature block is not used in memoranda.

EMAIL: FUNCTIONS, STYLE, FORMAT

Until the emergence of social networking services like LinkedIn and Facebook, email has seemed like the only communication tool you need for professional work. Despite their popularity, it's not likely that social networking services will replace email. The following focuses on email functions, style, and format.

Important Email Functions.   If you use email in professional work, make sure you are comfortable with the following not-so-basic skills:

- *Save email into files or folders.* Organize your sent and received email into meaningful folders—for example, "clients," "staff," "projects," "friends & family."
- *Keep copies of email you send.* Don't delete the email you send. Sometimes, it disappears into the Internet void, recipients may accidentally lose it, or you can't remember what you wrote.
- *Search email folders.* Know how to search your email folders for topics or for the names of recipients or senders. Inevitably, you'll forget what you wrote to a client or what that client wrote to you. Doing a quick search is far better than scanning through hundreds of emails.
- *Create and use distribution lists.* Increase your email efficiency by creating distribution lists (groups of related email addresses such as those for staff or clients).
- *Use a signature.* If you need to include your full name, title, organization, phone and fax numbers, and other such in your email, set up a "signature"—it automatically pops into every email you send.
- *Use templates.* If some of your email has standard content, set up templates. For example, create a template for your standard request for bids or announcement of services, import the template into your email, and just change the necessary details.
- *Proofread and spellcheck your email.* Spellcheck every email you send. Always proofread your email, checking specifically for missing words—which a spellcheck cannot catch. (Imagine the devastating effects of leaving out a single "not.")
- *Plan how to access your email on the road.* Know how to access email while on business trips: Use a satellite-based service that is always available everywhere. Otherwise, use a free email service such as Gmail; or use software that enables you to log into your office computer and work as if you were sitting right there.

Email Format and Style.   In the early days of email, all you had was plain text. Now you can include different fonts, different type sizes, additional colors, graphics, tables, and even animation in your email.

As for style in email messages, here are some suggestions:

- *Informality.* Adjust the tone of your email according to the recipients and situation. Informality is common in email, but think twice about using humor or sarcasm with business clients and higher-level management—especially those whose native language is not English.

- *Brevity.* Email messages are normally rather short—for example, under a dozen lines—and the paragraphs are short as well. Most people don't like extended reading on a computer screen. Consider putting lengthy messages in printable documents (formatted as PDF) attaching them to your email.

- *Specific subject lines.* To ensure email gets read and has the desired impact, make the subject line specific and compelling. If recipients have 60 to 70 messages waiting in the inbox and all they can see are the subject lines, they are more likely to read the ones that look important or interesting.

- *Important information first.* High-volume email users tend to lose interest or patience quickly. Put the most important information at the beginning of your message. Use subsequent sentences for elaboration.

- *Short paragraphs and space between paragraphs.* Whenever possible, break your messages into paragraphs of three or four lines. And when you divide your message into paragraphs, skip a line between them.

- *Highlighting and emphasis.* Use typographical effects (bold, italics, color, different fonts), tables, and graphics consistently and in moderation. You can also use tables and graphics to reinforce your messages (see Chapter 7).

- *Headings.* If your email requires readers to press the PageDown key, use headings to identify the subtopics within the message. For these headings, use a slightly larger font and bold. (See Chapter 3 for details on headings.)

- *Lists.* If you have key points to emphasize or if you are presenting step-by-step information, use bulleted and numbered lists, respectively. Contemporary email software provides automatic formatting for both types of lists. (See Chapter 3 for more on lists.)

- *Automatic replies.* The reply function in email is a great time saver—but also a disaster waiting to happen. Because email is often addressed to multiple recipients, it's easy to broadcast a reply to all of them when you intended to reply to only one. Imagine that in replying to a partner on a project, you question the competence of another partner on that same project. But what if this reply also goes to both partners? Uh oh...

- *Other concerns.* As mentioned early in this chapter, email has become so overused and overabused that our in-boxes are often overwhelmed. See the recommendations for when *not* to send email in "Email, Instant Messages, or Paper?" earlier in the chapter.

NEW INTERNET MEDIA

And God knows we need a better future for email, because the present is intolerable. This once-miraculous productivity tool has metastasized into one of the biggest timesucks in American life. Studies show that there are 77 billion corporate email messages sent every day, worldwide. By 2012, that number is expected to more than double. The Radicati Group calculates that we already spend nearly a fifth of our day dealing with these messages; imagine a few years down the road, when it takes up 40 percent of our time.

Clive Thompson, "The Great American Timesuck,"
Wired, July 2008, p. 58.

Clive Thompson, along with Dave Pollard mentioned earlier, are among many who lament the avalanche of email that professionals receive. As the following review of new Internet media shows, a number of alternatives are out there that can reduce this problem.

The twenty-first century has seen the rise in popularity of new Internet media such as blogs, wikis, screen-sharing software, and social-networking services. While some of these new media may seem like fads for teenagers, professional engineers are increasingly making use of them. These new media provide extra dimensions to the way professionals can communicate and work together.

> **Knee power**
>
> Engineers at several British universities have developed a wearable piezoelectric device that converts knee movement into electricity, which could in turn be used to power gadgets such as heart rate monitors, pedometers, accelerometers, and mobile phones.
>
> For details, see the Preface for the URL.

ADVANCED FORUMS

Online forums offer a more expansive approach to communicating with other engineers. The initial forums on the Internet were difficult to use and offered limited interactivity. However, forum applications now offer some useful functions:

- You post your question or comment as usual in any online forum.
- You provide your email address so that you get email notification when anyone responds to your forum post.
- You indicate topics of interest, and the forum application sends you email whenever those topics are contained in a forum post.

You can find engineering forums for just about every field of engineering by using "engineering forum" as a search term. See "Engineering Exchange" at www.engineeringexchange.com/forum. At engineersedge.com, for example, you would find discussions of optimum efficiency for flues in brick ovens and an inquiry about hydraulic seal.

Note   See the corresponding chapter in the website companion for steps in using an engineering forum. For the web address, see the Preface.

Blogs and the Blogosphere

The blog is another alternative to email. Fundamentally, it is an online journal that others can append comments to. A blog is what you are thinking about, what you're "into," what you are working on, and—frankly—anything else you are interested in. When people subscribe to your blog, they are notified through their blog aggregator if you post some topic in your blog that they are interested in. See "Top 100 Engineering blogs" at www.engineerjobs.com/blogs/top-engineering-blogs.php

See Chapter 12, "Engineering Your Online Reputation," for details on blogs.

Meeting and Screen-Sharing Software

Another way to get past problems with email involves meeting software, which typically incorporates "screen-sharing" and video and audio conferencing—all accessible through a simple web page. For example, if the meeting is focused on a drawing or a document, meeting attendees can display drawings or documents and the rest can comment on them.

Once again, what could have been a torrent of email exchanges can be handled neatly and efficiently. Figure 4-6 shows a screen capture of an Adobe WebConnect session. The current meeting attendee is waiting for others to show up so that they can discuss and edit the hardware graphic.

Note   See the corresponding chapter in the website companion for links and startup information for meeting applications. For the web address, see the Preface.

Social-Networking Software

One final instance of new Internet media that can replace the overreliance on email involves what is currently called social-networking websites. Examples like Facebook.com combine both fun, personal communications as well as professional communications. Examples like LinkedIn.com seem more for professional communications solely.

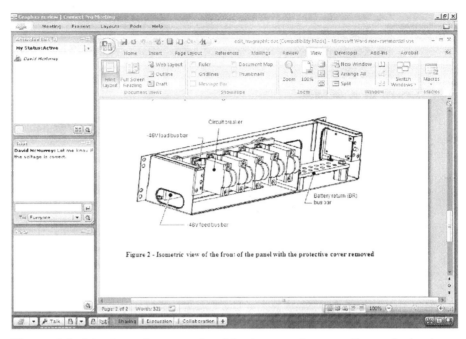

Figure 4-6 Meeting software session. Meeting attendees can discuss the hardware graphic, either using audio or the Chat pod. The meeting host (the only one currently present) would probably make the changes agreed upon by the attendees.

A presence at LinkedIn.com offers you a place to post your résumé, a listing of your projects, a place where you can explore your thoughts in blog form, a place for "friends" to make comments, and other such features.

See Chapter 12, "Engineering Your Online Reputation," for details on social networking sites.

EXERCISES

Talk to several professional engineers about the business correspondence they write or receive:

1. What are the typical audience, purpose, and content of their letters and memos? Why letters and memos as opposed to phone calls?

2. How much secretarial assistance do they receive? Do they get any help editing or proofreading their correspondence?

3. When they have to convey specialized, technical information, is it to another engineer, or must they often translate for nonspecialists?

4. How do they decide between writing a hardcopy letter or memo, making a phone call, or sending email?

5. What do they see as the advantages and the problems of using email in conducting their business?

6. Do they use some form of social media? Which ones and how?

BIBLIOGRAPHY

Chan, J.F. *E-Mail: A Write It Well Guide—How to Write and Manage E-Mail in the Workplace.* Write It Well, 2008.

Flynn, N., and Kahn, R. *E-Mail Rules: A Business Guide to Managing Policies, Security, and Legal Issues for E-Mail and Digital Communication.* AMACOM, 2003.

Hay, D. *A Survival Guide to Social Media and Web 2.0 Optimization.* Wiggy Press, 2009.

Hirsch, H. *Essential Communication Strategies: For Scientists, Engineers, and Technology Professionals.* Wiley–IEEE Press, 2002.

Kabani, S. and Brogan, C. *The Zen of Social Media Marketing: An Easier Way to Build Credibility, Generate Buzz, and Increase Revenue.* BenBella Books, 2012.

Lindsell-Roberts, S. *Strategic Business Letters and E-mail.* Houghton Mifflin, 2009.

Perry, T.S., and J.A. Adam. Email Pervasive and Persuasive. *IEEE Spectrum* (October 1992): 22–23.

Pollard, D. "When Not to Use Email." (February 6, 2007). http://howtosavetheworld.ca/2007/02/06/when-not-to-use-e-mail/. Accessed February 6, 2012.

Sahn, W.A. *The Gregg Reference Manual.* McGraw-Hill, 2010.

5

WRITING COMMON ENGINEERING DOCUMENTS

This project requires as much writing as it does engineering!

> Lev Shuhatovich, Engineer, Appliance Lab LLC, Austin,
> Texas, 2008.

This chapter explores common types of documents you may write as an engineer, focusing on their typical content, organization, and format. As you read this chapter, keep in mind that the names of these types vary considerably, and their contents often combine in different ways:

1. **Inspection or trip reports.** Briefly report on the inspection of a site, facility, or property; summarize a business trip; or report on an accident, describing the problem, discussing the causes and effects, and explaining how it can be avoided.

2. **Research, laboratory, and field reports.** Report on an experiment, test, or survey; discuss the research theory, method, or procedure; present the data collected; discuss conclusions, and possibly, explore applications of the findings or possibilities for further research.

3. **Specifications.** Provide detailed requirements for a product to be developed or detailed descriptions of an existing product; provide specifics on design, function, operation, and construction.

4. **Proposals.** Seek a contract, approval, or funding to do a project; function as a competitive bid to get hired to do a project; promote you or your organization as a good candidate for a project; promote the project itself, showing why it is needed.

5. **Progress reports.** Summarize how your project is going, what you or your group has accomplished, what work lies ahead, what resources have been used, what problems have arisen.

6. **Instructions.** Explain how to perform tasks, provide procedures on using equipment, give troubleshooting and maintenance guidelines, explain policies and procedures.

7. **Recommendation and feasibility reports.** Study a situation or problem, report on various alternatives, recommend the best one, or assess the feasibility of a project.

The documents discussed in this chapter are mostly short and informal and are routinely formatted as in-office memoranda or business letters. However, practically any of these documents can be formatted as full-length formal reports. (See Chapter 6 for the design of full-length reports.)

> **Solar shingle power**
>
> Dow has developed solar shingles that combine the functions of both weatherproof shingles *and* solar panels in one unit. They can be installed like typical roofing materials.
>
> For details, see the Preface for the URL.

Note   See the corresponding chapter in the website companion for examples of these reports. For the web address, see the Preface.

SOME PRELIMINARIES

Before getting into the details, consider some points that apply to all the types of documents about to be discussed:

- **Don't obsess over the names of reports.** Sorry, there is no ANSI standard on the proper names, contents, and format of reports. Don't worry about whether a document is really an evaluation report or really a recommendation report. Determine the requirements for the document you must write. One or some combination of the report types discussed here is likely to suffice.

- **Find out your company's requirements.** This chapter illustrates common characteristics of these documents. However, every company, organization, field, and profession has its own names and expectations about these documents. Those expectations may be written somewhere, or something everybody "just knows." Your job is to find out what those expectations are. The discussion and examples in this chapter give you some clues about what to expect and something to use when there are no guidelines.

- **Think creatively about content and organization.** Rarely will the contents and organization of the documents described here be a perfect fit for your real-world projects. The plans presented here cannot always be used as templates. Always think creatively, brainstorming about what else your readers may need.

- **Build your documents on the needs of your audience.** Everything about your documents depends on the specific people who are going to read it. Sometimes you must write for different audiences within the same report. See Chapter 3 for detailed discussion of analyzing and adapting to audiences.

- **Be careful with discussion of technical background.** Background sections provide information only to make the rest of the document understandable. Loosely related background sections are not helpful. Write the main text of your documents first, then review it for what readers may need help with; only afterwards write the background section, with the readers' needs in mind.

- **Be careful with the report introduction.** An introduction introduces readers *to the document*—not to the technical subject matter. The introduction announces the topic, alludes to the situation that caused the need for the document, indicates required audience knowledge, and provides a brief overview of the topics to be covered (and not covered). It's bad practice to dive right into the main subject matter in an introduction—readers then lack any perspective, overview, or roadmap for the whole document.

- **Packaging.** These documents can be presented as memoranda if they are short and addressed internally, or they can be presented as formal reports, with covers, table of contents, and appendixes. For reports over three or four pages, consider using the formal-report format shown in Chapter 6.

- **Format.** Just because it is a short and informal report, don't neglect to use basic formatting practices that will make your report more readable, more usable, and more accessible—not to mention more professional in appearance:

  - Unless the report goes over several pages or unless your company has certain requirements, use the memorandum format. (If you are writing to someone external to your company, use the business-letter format.)

  - Use headings to mark off the major subtopics within the report. Notice how they are used in the example report in Figure 5-1. Headings help readers skip to the sections they want to read.

  - Use the various types of lists as needed. These help emphasize key points, make information easier to follow, help readers return to key points, and generally create more white space—all of which makes your report more readable.

- **Media.** If you are expected to produce a portable document file of your report, see the section in Chapter 6, "Generating Portable Document Files," for some recommendations. See Chapter 12, "Engineering Your Online Reputation," for ways you can team-write this type of report and make it available for review and comment.

Note   Additional examples are available at the companion website for this book. See the Preface for the web address.

INSPECTION AND TRIP REPORTS

One common group of engineering documents handles tasks such as those discussed in the following. These types of documents are referred to in various ways and they overlap considerably.

- **Trip reports.** Discuss the events, findings, and other aspects of a business trip. This type reports your observations so that people in your organization can share them (see Figure 5-1).

- **Investigation or accident reports.** Describe your findings concerning a problem; explore its causes, its consequences, and what can be done to avoid it.

- **Inspection or site reports.** Report your observations of a facility, a property, or an installation of equipment, with description and possibly evaluation of it.

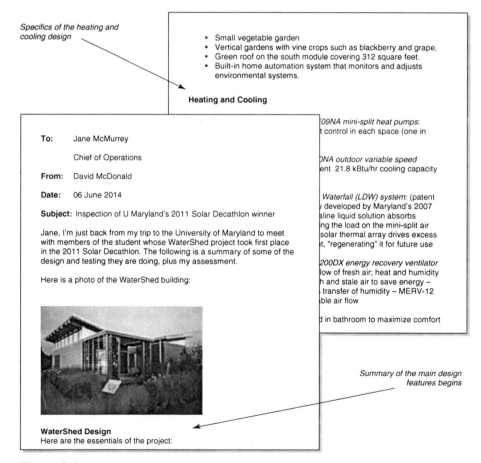

Figure 5-1 Short business-trip report—excerpts.
Source: Adapted from www.gizmag.com/2011-solar-decathlon-winner/20104

For the content of an informal report, consider these suggestions:

INTRODUCTION

No matter which type of report you write, begin by indicating the purpose of the report and providing a brief overview of its contents. Avoid that impulse to dive right into the thick of the discussion!

BACKGROUND

It's also a good idea to set the stage—to explain the background or context of the report. Why did you go on this business trip? Why were you sent to inspect the facility? Who sent you? What are the basic facts of the situation—the time, date, place, and so on?

FACTUAL DISCUSSION

The main contents of reports like these are some combination of description, narration, or both. Typically, you must describe the accident, facility, property, or the proposed equipment. You must also give a narrative account of what happened on the trip: where you went; who you met with; what was discussed.

ACTIONS TAKEN

If you are investigating a problem and implementing a solution, provide a step-by-step discussion of how you determined the problem and corrected it.

INTERPRETIVE, EVALUATIVE, OR ADVISORY DISCUSSION

Once you've laid the foundation with the background and factual discussion, you're set to do what readers may expect—to evaluate the property or equipment, explain what caused the accident, interpret the findings, suggest further action, or recommend ways to prevent the problem in the future.

RESEARCH, LABORATORY, AND FIELD REPORTS

Laboratory and field reports present not only the data from an experiment or survey and the conclusions that can be drawn from that data, but also the theory, method, procedure, and equipment used. (See the excerpts from a laboratory report in Figure 5-2.)

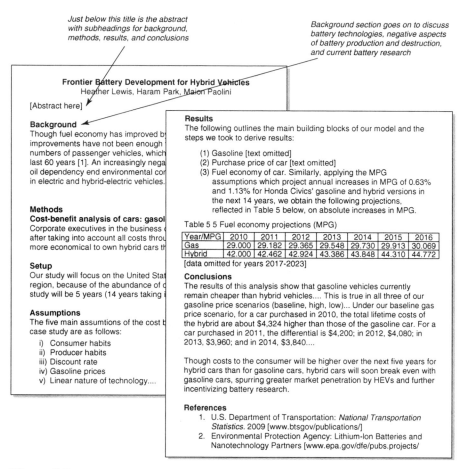

Figure 5-2 Excerpts from a primary research report. Lewis et al., *Chemistry Central Journal* 2012, 6(Suppl 1):52 [http://journal.chemistrycentral.com/content/6/51/53].

Note Many resources point to this type of report as *the* engineering report. See the format for the formal version of this report in Chapter 6.

As much as practical, the laboratory or field report should enable readers to replicate the experiment so that they can verify the results for themselves. Because of this dual requirement, laboratory and field reports have a characteristic structure.

INTRODUCTION

Indicate the overall topic and purpose of the report, and provide an overview of its contents. Remember: Avoid diving into the thick of the discussion; instead, orient readers to the report topic, purpose, and contents first.

BACKGROUND

Provide a discussion of the background of the project. Typically, this involves discussing a research question or conflicting theories in the research literature. Or, for example, apply an interesting discovery from another field to something in your own. Explore this background to enable readers to understand why you are doing this work. When you do, provide citations for the sources of information you use, using the standard bibliographic format (see Chapter 11).

LITERATURE REVIEW

Usually included in a lab or field report is a discussion of the research literature related to your project. You summarize the findings of other researchers that have a bearing on your work. Again, use the standard bibliographic format.

Depending on the length and complexity of the report, all three of the elements just discussed—introduction, background, and literature review—may easily combine into one introductory paragraph; or they may occur separately under their own headings. Regardless of their length, these three elements should occur at the beginning of a laboratory or field report, even if each one is brief.

THEORY, METHOD, PROCEDURE, AND EQUIPMENT

Next, present your theory or approach to the project. For example, as a software engineer, you may suspect that computer users prefer social media tools to traditional email. To test this idea, you create a questionnaire and send it to a cross-section of computer users. In this report, you'd first discuss the common thinking on this subject. Then you'd explain your method and procedures. Readers could use this part of the report in particular to replicate your project.

OBSERVATIONS, DATA, FINDINGS, AND RESULTS

Collect data and then organize and present it in a section of its own. The common approach is to present the data, often formatted into tables, graphs, or charts, without interpretive discussion.

CONCLUSIONS

In the conclusion section, you draw conclusions based on the data you've gathered and explain why you think those conclusions are valid.

IMPLICATIONS AND FURTHER RESEARCH

Laboratory and field reports also typically explore the implications of conclusions, considering how they can be applied and outlining further research possibilities.

These three sections—findings, conclusions, and implications—can be rolled into one if the report is brief and relatively simple; or they can occur in sections of their own with headings. In whatever way they are combined, the first two elements—the data and the conclusions—must occur.

INFORMATION SOURCES

Conclude with a section that lists information sources used in the project. For entries in that list, use the bibliographic format shown in Chapter 11.

SPECIFICATIONS

Specifications are standards and requirements relating to products, services, and even of materials. They provide details for design, manufacture, testing, installation, performance, and use. Be aware that some controversy exists over the distinction between specifications and requirements. One type of specification called the data sheet lists specifications primarily in table format with little extended writing. See Figure 5-3 for an example.

The type of specifications here focus on construction, operational, and perfomance details. Accuracy, precision of detail, and clarity are critical. Poorly written specifications can cause a range of problems including lawsuits.

Specifications have a particular style, format, and organization. If you write specifications, find out what those are in your organization or field. If these are not documented, collect specifications written for your company or field, and study them. Here are some general recommendations:

- In table-style specifications (called data sheets), use two-column lists or tables (as shown in Figure 5-3) to list specific details. Because the purpose is to indicate details such as dimensions, materials, weight, tolerances, and frequencies, regular paragraph-style writing is minimal.
- Make sure each specific requirement is separate and uses the decimal numbering system for ease of cross-referencing.
- Begin with information about the author, authorities, and dates of the specification, as shown in Figure 5-4.
- Use the standard introductory elements after the opening section. Figure 5-5 shows an example.
- Use a standard organization of specifications such as the one shown in Figure 5-6.
- In the sentence-style format, use an outline style similar to the one shown in Figure 5-7. Each specification must receive its own number-letter designation so that each can be referenced separately.
- Use one of two writing styles, depending on the requirements of the job. In the *open* or *performance* style, specify what the product or component should do or what should be done—that is, its performance capabilities (see Figure 5-7). In the *closed* or *restrictive* style, specify exactly what it should be or consist of (see Figure 5-3).

Centaurus II Specifications	
Dimenstions	
Length	5 m
Width	1.8 m
Height	1 m
Weight	176.5 kg
Motor	
Type	NGM SMC150
Rating	7.5 kW
Array	
Cells	China Sunergy
Peak Power	1200 Watts
Battery	
Type	Lithium-Polyn
Weight	25 kg
Capacity	2.8 kW-hour
Body	
Chassis	Fiberglass co monocoque v
Brakes	
Front	Redundant, h
Rear	Regenerative
Wheels	
Front	NGM Alumini
Rear	Custom
Tires	9-ich Bridges
Suspension	
Front	Unequal-leng
Rear	Swing arm

MIL-STD-1553B DATA BUS REQUIREMENTS

COMMUNICATIONS LINE:

Cable Type	Two-conductor twisted pair
Capacitance	30 Pf/ft., max.
Twist	Four per foot, min.
Char. Impedance (Zo)	70 to 85 ohms at 1 MHz
Attenuation	1.5 db/100 ft @ 1MHz max.
Bus Length	Not specified
Termination	Two ends terminated in resistors = (Zo) ± 2%
Shielding	90% coverage min., 00% dual standby redundant

CABLE COUPLING:

Stub Length	Up to 20 feet (may be exceeded)
Stub Voltage	1-14V p-p amplitude, line-to-line min.
	Signal voltage, transformer coupled

COUPLER TRANSFORMER:

Tums Ratio	1.41:1
Droop	<20% (1)
Overshoot/Ringing	<+1V (1)
CMR	>45 dB at 1 MHz (1)
Fault Protection	Series resistors = 0.75 Zo ± 2%

NOTES:
(1) at 27 V p-p 250kHz square wave. CMR = Common Mode Rejection

Figure 5-3 Data sheet type for milspec specifications and for Centaurus II (University of Minnesota's 9th-generation solar vehicle, which took 2nd place in the 2010 American Solar Challenge).
Source: www.milestek1553.com and www.umnsvp.org/?page=vehiclesC2

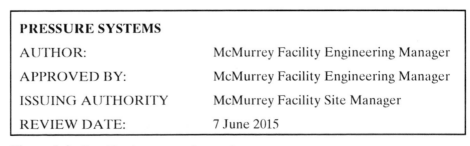

PRESSURE SYSTEMS	
AUTHOR:	McMurrey Facility Engineering Manager
APPROVED BY:	McMurrey Facility Engineering Manager
ISSUING AUTHORITY	McMurrey Facility Site Manager
REVIEW DATE:	7 June 2015

Figure 5-4 Specifications—opening section.

MODIFICATION OF BUILDINGS AND BUILDING SERVICES

1. PURPOSE

To ensure that when buildings or building services are modified it is done with the approval of the McMurrey Facility Site Manager, at the appropriate technical standard, and in compliance with current legislation. Also, ensure that relevant records are updated.

2. SCOPE

This instruction applies to all such modification work in the McMurrey Facility Area.

3. It does not apply to modification of research equipment, to which the standards of Ref. 3.1 apply.

4. REFERENCES

4.1 ASG/WCE/2007: Control of Modifications to Plant

Figure 5-5   Specifications—introductory elements.

1. PURPOSE
2. SCOPE
3. REFERENCES
4. DEFINITIONS
5. INSTRUCTION
5.1 DESIGN
    5.1.1 Identification of Pressure Systems
    5.1.2 Pressure Vessels
    5.1.3 Relief Streams
5.2 REGISTRATION
    5.2.1 Pressure Vessels
    5.2.2 Relief Streams
5.3 DOCUMENTATION
    5.3.1 Pressure Vessels
    5.3.2 Relief Streams

Figure 5-6   Specifications—common outline (body text omitted).

PART 1 EXECUTION

1.1 EXAMINATION

A. Do not begin installation until substrates have been properly prepared.

B. Verify mounting surfaces are ready to receive fixtures.

C. If substrate preparation is the responsibility of another installer, notify the architect of unsatisfactory preparation before proceeding.

1.2 PREPARATION

A. Clean surfaces thoroughly prior to installation.

B. Prepare surfaces using the methods recommended by the manufacturer for achieving the best result for the substrate under the project conditions.

1.3 INSTALLATION

A. Install in accordance with manufacturer's instructions.

B. Install surface-mounted emergency lighting units plumb and adjust to align with building lines and with each other. Secure these units to prevent movement.

C. Connect emergency lighting units to branch circuit outlets provided as indicated in the drawings.

D. Install specified lamps in each emergency lighting unit.

Figure 5-7 Specifications—writing style and format.

- Whenever possible, cross-reference existing specifications rather than repeating those details. Government agencies as well as trade and professional associations publish specifications standards. Refer to these standards rather than copying them verbatim into your own specifications.

- Use specific, concrete language that identifies as precisely as possible what the product or component should be or do. Avoid ambiguity (using words that can be interpreted in more than one way). Use technical jargon the way it is used in the trade or profession.

- A traditional writing style for specifications has been to use *shall* to indicate requirements. In specifications writing, *shall* is understood as indicating a requirement. More recently, simple imperatives are being used as shown in Figure 5-7. (In the traditional style, 1.3.D would be "Specified lamps shall be installed in each emergency lighting unit.")

- Provide numerical specifications in both words and symbols: for example, "the distance between the two components shall be three centimeters (3 cm)."

- Use a relatively terse writing style in specifications. Incomplete sentences are acceptable as well as the omission of obvious function words such as articles.

- Exercise caution with pronouns and relational or qualifying phrases. There should be no doubt about the reference of words such as *it, they, which*, and *that*.

Watch out for sentences containing a list of two or more items followed by some descriptive phrase—does the descriptive phrase refer to all the list items or just one? In these cases, use a wordier approach for the sake of clarity.

- Use words and phrases that have become standard in similar specifications over the years. Past usage has proven them reliable. Avoid words and phrases that are known not to hold up in lawsuits. (This is one reason why it is wise to work with an experienced specifications writer.)

- Make sure your specifications are complete. Put yourself in the place of those who need your specifications; make sure you cover everything they will need.

Test your specifications by putting yourself in the role of a bumbling contractor—or even an unscrupulous one. What are the ways a careless or incompetent individual could misread your specifications? Could someone willfully misread your specifications in order to cut cost or time? Obviously, no set of specifications can ultimately be "foolproof" or "shark-proof," but you must try to make them as clear and unambiguous as possible.

> **Solar paint power**
>
> Scientists at USC are developing a technology to cheaply produce stable liquid solar cells that can be painted or printed onto clear surfaces. Using this technology, cheap, flexible solar panels could be shaped to fit just about anywhere.
>
> For details, see the Preface for the URL.

PROPOSALS

An important tool, particularly if you are a consulting engineer, is the proposal. With it, you get work, for either your company or yourself.

Proposals are defined in different ways. In this book, however, the proposal is something quite specific: It is a bid, offer, or request to do a project plus supporting information necessary to gain approval to do the project. Proposals sometimes must convince the recipient that the project needs to be done, but proposals must always convince the recipient that the proposer is the right individual or organization to do the project.

In the formal proposal scenario, an organization sends out a request for proposals (RFP) to do a project. An RFP might be published in newspapers, professional journals, or specialized periodicals such as the *Commerce Business Daily*; sent by mail to a select list of vendor organizations; or conveyed by various informal means such as telephone or email. The organization receiving the RFP then writes and submits a proposal presenting its qualifications and making a case for itself as a good choice. The recipient of the proposals selects one of the proposals and enters into contract negotiations. Once that is accomplished, the organization that wins the project can get down to work.

Proposal writing is a competitive affair. Highlight your organization's strengths; make a good case for your company as the right one for the project.

TYPES OF PROPOSALS

Proposals are commonly divided into two types, based on whether the recipient requested them:

- **Solicited.** If an organization issues a request for proposals, the proposals are "solicited"—they have been requested.
- **Unsolicited.** Individuals and companies often initiate proposals without a formal request from the recipients. They may see that an individual or organization has a problem or opportunity. When the proposal is unsolicited, you, the proposal writer, have to do the additional work of convincing the recipient that the project needs to be done.

Proposals can also be divided according to the context in which they occur:

- **Internal.** If you address your proposal to someone within your organization, format and contents change significantly. The memo format is usually appropriate, and background and qualifications may not be necessary.
- **External.** If you address your proposal to an organization outside of your own, you must present your qualifications and use some combination of the business-letter and formal-report formats.

In some sources, proposals are "great idea" documents such as developing highway-integrated wireless-charging technology. Great idea—just by driving your car you charge its batteries! But other resources would define this great-idea document as recommendation report or an engineering design report. What would make this document a proposal are the components described in the following.

The typical sections in a proposal are as follows (see the proposal excerpts in Figure 5-8 in which some of these sections are illustrated):

Introduction.   In the first section of a proposal, refer to prior contact with the recipient of the proposal or your source of information about the project. Identify the information that follows as a proposal (in other words, state the purpose). Also, briefly overview the contents of the proposal.

Background.   In an unsolicited proposal, discuss the problem or opportunity that caused you to write the proposal. In solicited proposals, this may not be necessary: The party requesting proposals knows the problem very well. Still, a background section even in a solicited proposal can be useful: It enables recipients to check your understanding of the situation.

Actual proposal statement.   Include a short section stating explicitly what you propose to do. Proposals often refer to many possibilities, which can create some vagueness about what's actually offered. You may also need a scope statement—an explicit statement about what you are *not* offering to do.

To: David A. McMurrey
            Proposal approval instructor
From: Miriam Ximenez, Geotechnical engineer
Date: May 25, 2013
Subj: Proposal to develop an informative report on main probabilistic methods to
            evaluate uncertainty in geotechnical engineering

As my last email sent on April 6 promised, this proposal presents a plan to develop a
report that explains importance of evaluating uncertainty in geotechnical engineering
using probabilistic methods. I will address your concern that language used in the
report may become too technical for clients to understand. Technical sections will
address basic probabilistic concepts and will be more oriented to real cases in which
traditional methods are more costly than probabilistic methods.

Background

Not all projects call for the application of probabilistic methods. If site conditions for a
project are homogeneous, conventional procedures can be used instead. However, if site
conditions are heterogeneous, a probabilistic approach is in order. Similarly, for small
projects, conventional pro
probability of failure is gre

Proposal

This proposal seeks appr
analysis and persuades c
probabilistic method and
advantages of probabilist

Benefits: Explaining the
clients to use this type of
probabilistic methods can
design parameters.

Description of the Repo

Here are some specifics
Technical background.
methods applied to geote
method.

Report outline. The mair
comparison tables. For e
most appropriate probabi
report:

   I.   *Introduction.* Su
   II. *Background on pr*
       applied to calculat
        A. Site evalua
        B. Variability c
        C. Reability a
          parameters

   III. *Applications of probabilistic methods.* Two areas in which probabilistic methods
       are useful:
        A. Risk evaluation of dams
        B. Risk projections involving earthquakes
   IV. *Cost comparisons.* Reports with and without probabilistic analysis
   V. Conclusions. Review of the accuracy and cost advantages of probability methods
       and their limitations

Schedule.   The schedule for developing this report is as follows:

Proposed start of project	May 4, 2013
Completion of research and gathering of material for the report	May 15, 2013
Report writing and preparation of graphics	May 25, 2013
Editing and proofreading of report	May 30, 2013
Delivery of completed report	June 6, 2013
Revisions as requested by your review	June 15, 2013

Qualifications. My qualifications for this work are as follows:

* Risk analysis studies in my geotechnical engineering masters at Virginia Tech
* Thesis on probabilistic methods as applied to liquefaction analysis
* Five years experience as a geotechnical engineer (see attached résumé).

Cost. The total in-house costs are $3,100.00. This assuming 4 hours of writing time per
page at $50.00 per hour. It does not include revision and changes after revision. Editing,
and graphics production is estimated below.

Minimal research (4 hrs @ $200/hr)	800.00
Writing (10 pgs @ 4 hrs/pg @ $50.00/hr)	2,000.00
Editing, graphics, production	300.00
Revisions as requested (estimated, at writing rate)	250.00
TOTAL	$3,350.00

Conclusions. Production of an informative report will inform our clients about how
probabilistic analyses can be used to evaluate uncertainty in geotechnical engineering.

I look forward to discussing this proposal with you and to begining the project.

Figure 5-8   Proposal excerpts. Notice that this internal proposal (in memorandum
format) still estimates expenses for the project.

Description of the work product.   Some proposals need a section in which the
proposed project—in other words, the results of the work—is described. This might
be a constructed building, a program design, blueprints or plans, or even a 40-page
report. The point is to provide details on what the recipients will get.

Benefits and feasibility of the project.   To promote the project to the recipient, you
may need to discuss the benefits of doing the project and perhaps even the likelihood

of those benefits. This is particularly true in unsolicited proposals where the recipient must be convinced that the project is necessary in the first place.

Method or approach.   You may need to include a section explaining how you plan to go about the project, perhaps even the theory relating to your approach. For some projects, people need to know how the work will be done and why it will be done that way. As in the background section, this discussion enables you to demonstrate your professional expertise.

Qualifications and references.   Most proposals list the proposing organization's key qualifications, along with references to past work. This section is like a mini résumé. Large proposals actually include full résumés of the individuals who will work on the project.

Schedule.   Include a schedule of the projected work with dates or a timeline for the major milestones. Again, this gives the recipient an idea of what lies ahead and a chance to ask for changes; and it enables you to show how systematic, organized, and professional you are.

Costs.   Some proposals have a costs section that details the various expenses involved in the project. Rather than toss out a lump sum, break it into different kinds of labor, hourly rates for each, and other charges. If you are writing an internal proposal, you may need to list supplies needed, expenses for new equipment, your time (even though it is not charged), and so on.

Conclusion.   In the final paragraph, urge the recipients to consider your proposal, contact you with questions, and of course accept your bid or request. This is also a good spot to allude once more to the benefits of doing the project.

A fascinating example of a proposal written by engineering students at Frank W. Olin School of Engineering focuses on deployable greenhouse design for the planet Mars: http://projects.olin.edu/marsport

PROGRESS REPORTS

Another common report type is variously called the progress report, status report, interim report, quarterly report, or monthly report. Its job is to present to your clients or supervisors the status of your work. They can then act as manager of the project, request changes if necessary. In this situation, you are the supplier of the work of the project; the recipient of the work is the client—even internally.

A phone call to your client to provide an update on progress might be tempting. Bad idea. There will be no permanent record of any concerns that either your client or you raised. In most projects, changes, new requirements, problems, and miscommunications are bound to occur. Clients may worry that the work is not being done properly, is not on schedule, or is not within budget. Suppliers, on the other hand, may worry that clients will not like how the project is developing, that new requirements jeopardize the schedule and budget for the project, or that unexpected problems have arisen. Therefore, the progress report can allay clients' concerns and can help suppliers stay in touch with their clients and protect themselves from unreasonable expectations and unwarranted accusations.

Because of these functions and expectations, progress reports typically have the following contents and organization (see the excerpts from a progress report in Figure 5-9):

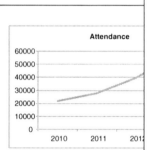

Date: August 8, 2013

LCRA

To: Jim Travers
From: Joanne Riceland, project assistant
Subj.:   Status of the Osprey Point Hall project at
Lake Bastrop South Shore Park

The following is a report on the progress of improvements to Osprey Point Hall at Lake Bastrop South Shore Park. Included in this report is an overview of the project and discussion of work completed, work remaining and recommendations.

PROJECT OVERVIEW

Osprey Point Hall, an enclosed pavilion, was built to provide a basic park amenity that would benefit the Parks Department and the local economy and attract visitors. In addition, rental of the hall is expected to increase annual revenue by $37,000 per year. For a capital investment of $575,000, the completed project is expected to meet the goals of enhancing amenities for park visitors and helping to meet the department's 50% park system cost recovery goals.

Attendance at Lake Bastrop South Shore Park has almost doubled since 2010 as demonstrated in Figure 1. Increasingly, park visitors have been requesting a simple enclosed structure with heating, air-conditioning and a food preparation area that can be used for meetings, weddings and family gatherings.

Attendance

60000
50000
40000
30000
20000
10000
0
      2010   2011   201

Figure 2. Lake Bastrop South Shore Park

The capital project was approved in January
July 2012. The pavilion was certified for occ
currently is in the close-out phase with an e
September 2013.

WORK COMPLETED

Osprey Point Hall (Fig. 2) was certified for o
structure met all ADA inspection standards f
other permitted requirements. The native lar
is complete. Staff has purchased tables and
chairs for the porch. Several groups have re
and it is reserved at least twice weekly throu

WORK REMAINING

The project currently is in the close-out phas
Cavendar received the engineering drawings from the contractor, and will submit them to Records and Archives for permanent storage. Ms. Cavendar has scheduled a session to identify "lessons learned" that can be applied to future projects. Ms. Cavendish will begin the financial closeout once all final invoices have been paid. The project is expected to be completed within its lifetime budget target. Figure 3 shows the fiscal year expenditures by quarter toward the lifetime project budget of $575,000.

Figure 5-9   Progress report—first pages. If your progress report is short, you can incorporate the report into a business letter (or memo if it's internal), making it one continuous document as is done here.

INTRODUCTION

As with any report, start with the purpose and topic of the report, its intended audience, and a brief overview of the report's contents before diving into the thick of the discussion.

PROJECT DESCRIPTION

Briefly describe the project; some of your readers may not be familiar with it. Summarize details such as purpose and scope of the project, project start and end dates, and names of suppliers and clients involved. Unless the project changes, this description can become boilerplate text in future progress reports and appear under its own heading, enabling readers to skip it.

PROGRESS SUMMARY

The real substance of the progress report is the discussion of what work you've completed, what work is in progress, and what's yet to come. This discussion can be handled several ways:

- *Time-periods approach.* Summarizes work completed in the previous period, work underway in the current period, and work planned for future periods.
- *Project-tasks approach.* Summarizes which tasks in the project have been completed, which tasks are currently underway, and which tasks are planned for the future.
- *Combined approach.* Combines these approaches by dividing the section on previous-period work into summaries of the work done on individual tasks, or by dividing the project-task sections into summaries of work completed, in progress, or planned for each task.

Use whichever of these approaches works best in terms of your project and your client. For simpler projects, however, the time-periods approach works best. The project-tasks approach works well when the project has a number of semi-independent tasks on which you are working more or less concurrently.

PROBLEMS ENCOUNTERED

In this section, you go on record about the problems that have arisen in the project—problems you think may jeopardize the quality, cost, or schedule of the project.

CHANGES IN REQUIREMENTS

In this section, keep a history of changes in the project as you understand them: for example, schedule shifts and new requirements.

OVERALL ASSESSMENT OF THE PROJECT

In what is often the final section of the progress report, give an opinion as to how the project is going. In this section, resist the temptations to say that everything's going along just fine or to whine about every minor annoyance. Remember your job is to provide your clients with the details they need to act as managers or executives of the project as a whole. (See Figure 5-10.)

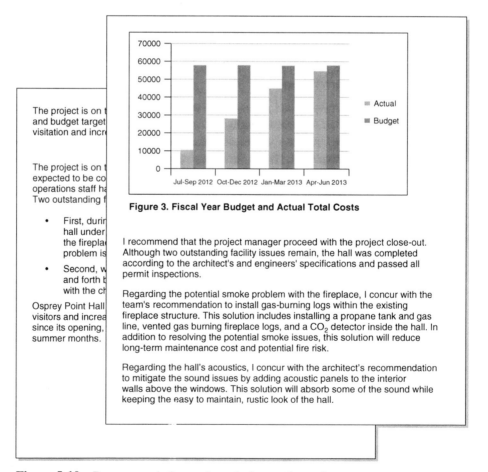

The project is on t
and budget target
visitation and incr

The project is on t
expected to be co
operations staff ha
Two outstanding f

- First, durir
  hall under
  the firepla
  problem is

- Second, w
  and forth b
  with the ch

Osprey Point Hall
visitors and increa
since its opening,
summer months.

Figure 3. Fiscal Year Budget and Actual Total Costs

I recommend that the project manager proceed with the project close-out. Although two outstanding facility issues remain, the hall was completed according to the architect's and engineers' specifications and passed all permit inspections.

Regarding the potential smoke problem with the fireplace, I concur with the team's recommendation to install gas-burning logs within the existing fireplace structure. This solution includes installing a propane tank and gas line, vented gas burning fireplace logs, and a CO_2 detector inside the hall. In addition to resolving the potential smoke issues, this solution will reduce long-term maintenance cost and potential fire risk.

Regarding the hall's acoustics, I concur with the architect's recommendation to mitigate the sound issues by adding acoustic panels to the interior walls above the windows. This solution will absorb some of the sound while keeping the easy to maintain, rustic look of the hall.

Figure 5-10   Recommendation and conclusion sections of a progress report.

OTHER SECTIONS

Other information may be required: for example, a summary of financial data on the project or the results of product testing. Be alert to the needs of your audience. If you're not sure whether progress reports are required, especially for short projects, check with your supervisor or client. While a "progress report" may be nothing more than a quick email, keep in mind that progress reports, when appropriate, strengthen your professional image. They keep you closer to your client and help eliminate unfortunate surprises.

Unless a schedule for progress reports is established by your supervisors or client, your sense of the project and the requirements of the client should dictate the number and frequency of progress reports. Typically, progress reports are sent at the end of every month or every quarter. The larger the project, the more formally defined these requirements are and the more formal the progress reports are.

> **Nanoantenna power**
>
> Researchers at Tel Aviv University have developed a system that converts optical waves by means of nanoantennas—very, very short antennas—that could replace the silicon semiconductors in special solar panels. This technology could harvest more energy from a wider spectrum of sunlight than is currently possible.
>
> For details, see the Preface for the URL.

Lengthy? See the Department of Energy's *FY 2002 Progress Report for Hydrogen, Fuel Cells, and Infrastructure Technologies Program* at www.eere.energy.gov/hydrogen andfuelcells/annual_report.html. It's 600 pages!

INSTRUCTIONS

In your engineering career, you may often find yourself writing step-by-step procedures for employees, colleagues, customers, or clients. In such instructions, you explain how to assemble, operate, or troubleshoot some new product your team is working on or how to operate equipment around the office, lab, or site.

> **Note**   If your job is to write a user guide, see the corresponding chapter in the website companion. For the web address, see the Preface.

SOME PRELIMINARIES

Critical in instructions writing is putting yourself in your readers' place, making no unwarranted assumptions about their background or knowledge, and providing them everything they need to successfully complete the procedure.

Understand the key difference between instructions and product specifications. In rushed development cycles, product specifications are sometimes used, with little revision, as instructions. That's unfortunate: Specifications do not function well as instructions. Specifications approach a product as a group of features and functions—not in terms of tasks. Consider your microwave oven: The statement "Power Cook button enables user to set power level" is specification language, not instructions. The user needs to know which other buttons to press and in what sequence. In specifications, the heading might be "Power Cook Function," whereas in instructions the heading would be something like "Cooking with Different Power Levels."

Critical in preparing to write instructions is audience analysis—identifying the relevant characteristics of the readers most likely to use your instructions. (For a full discussion of this task, see Chapter 3.) What do you expect your readers to know already, and what must you explain in your instructions? For example, in explaining how to install a computer program, you have to decide whether readers understand some basics about installation media, folders, directories, or files.

Introduction.   In the introduction, include some combination of the following:

- **Subject.** Indicate the procedure you'll explain. If you are providing instructions for a product, identify it.
- **Audience.** Indicate background your readers need in order to understand your instructions. If no special background is needed, indicate that.
- **Overview.** Briefly list the main contents of the instructions; for example, list the major tasks or procedures to be presented.

Special notices.   Most instructions contain specially formatted notices for warnings, cautions, and dangers. Often these appear in the introduction as well as in the body of the instructions at those points where they apply. If you neglect to include these special notices, you may find yourself in a lawsuit if readers injure themselves or lose money.

Style and format of special notices vary widely, but here's a recommended approach, which is used at IBM and other corporations:

- **Note.** For emphasizing special points or exceptions that might be overlooked.
- **Attention.** For alerting readers to a potential for ruining the outcome of the procedure or damaging the equipment.
- **Caution.** For alerting readers to the possibility of minor injury because of some existing condition (for example, the hazard of paper cuts when opening a ream of paper). Also used when a potentially dangerous situation might develop because of some unsafe practice (for example, making an unapproved hardware modification).
- **Danger.** For calling attention to a situation that is potentially lethal or extremely hazardous to people (for example, exposed high-voltage wires as a result of removing a computer side panel). Use this notice with discretion, reserving it for situations where irreparable injury or loss of life could occur unless extreme care is used.[1]

---

[1]Many thanks to Linda St.Clair, Editor, AIX & Eserver pSeries Information Design & Development, IBM Corporation, for updates on notices.

As you have read in the preceding pages, numerous components and formats get involved in instructions. They become only more involved in user guides and similar technical information.

Open some user guides listed in the technical support areas of major computer manufacturers:

- **Apple manuals.** Go to support.apple.com/manuals/Select user guides and other manuals for Apple products like iPad, Mac, iPhone, or iPod. Although Apple provides links to quick-start guides, they may not give you a full picture of user guides.

- **Dell manuals.** Try going to www.dell.com/support/Manuals/us/en/19/Product Selector/Select?rquery=na. You'll need to do some drilling here to get to actual manuals. For example, click on the Dell product named Inspiron, then click a link for a model. Set-up guides and service manuals are the best choices here to see a full range of components, formats, and styles as discussed in this part of the chapter.

- **Lenovo manuals.** Do the same at http://support.lenovo.com/en_US/guides-and-manuals/default.page?selector=expand. Take a look at the hardware maintenance manuals for the ThinkPad.

In the user guide for the iPad developed iOS 5.1, notice:

- How the headings are "hierarchical." For example, **Buttons** is a first-level heading, and **Sleep/Wake** and **Volume** are lower-level (second-level) headings. On the same page, notice that there are bold, black, "run-in" (third-level) headings: for example, **Set the Auto-lock time** and **Set a passcode**.

- How notices are used. On the same page with the **Volume buttons** heading, notice the warning. In Apple corporate style, warnings are used when any kind of physical harm—from a burn to electrocution—could result.

See Figure 5-11 for examples of these special notices. Serious ones are placed right at the point at which readers might wreck their procedure, ruin their equipment, hurt somebody, or blow themselves up! Standards for notices vary: What is a caution in one company or industry is a warning in another. ANSI defines a standard, but it is not widely understood or followed.

Background.   For certain complex tasks, readers need to know conceptually what they are doing. Positioned after the introduction and before the actual procedures, a background section enables readers to figure out much of the procedure, or its finer points, on their own. As a way of avoiding unnecessary background, write background only *after* you've written the procedures. A good example where background is necessary is the custom color feature found in graphics software. You simply cannot know what you are doing unless you have some background on hue, saturation, and other such concepts.

Operating the Industrial Scientific iTX Multi-Gas Monitor

The Industrial Scientific iTX Multi-Gas Monitor (iTX) is an adaptable portable gas detection instrument. The iTX can go from being a single-gas monitor to a six-gas monitor with just a few user configurations and sensor changes. Providing single-button operation, testing and calibration functions, the iTX's unique Quick-Cal feature calibrates up to four sensors at once, saving time and calibration gas costs.

Equipment and Parts

First, identify and layout all equipment and parts on a table:

- Industrial Scientific iTX Multi-Gas Monitor

Danger: Gas detection instruments are potential life-saving devices. Daily bump testing and proper calibration are essential.

- Calibration cup
- Polyurethane tubing
- Bump gas bottle
- Calibration gas bottle

Industrial Scientific iTX Multi-Gas Monitor

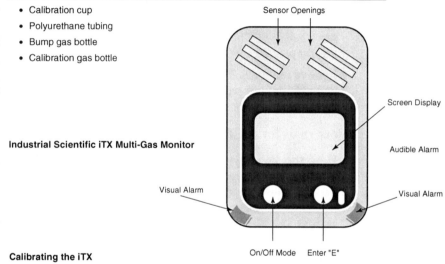

Calibrating the iTX

Caution: Make sure you use only Industrial Scientific-branded calibration gas.

1. Turn on the iTX and let it run through its start-up process.
2. Follow the bump test procedure to connect the monitor to the calibration gas.
3. Zero out the sensors next.
4. Press the ON/OFF MODE button until `Zero Sensors` shows on the monitor screen. Once all sensors have zeroed, the O2 CAL calibration screen will appear. The O2 CAL sensor will run through its calibration. If it fails, a replace sensor will show on the display screen. When zeroing process ends, the instrument will beep and display `Zero Complete` on the screen.
5. Keep pressing the E button until the `Calibration Yes` screen is displayed. Calibration will start and run through all sensors. `CAL in Progress` will display on the ...

Figure 5-11   Instructions—excerpts. Notice the use of the danger and caution notices as well as the use of Courier New for displayed messages. Example drawn from work done by Grantt Bedford, BP Safety Engineer.

Equipment and supplies. List the supplies and equipment that readers must gather before they begin. Supplies are consumable items used in the process. Equipment is the tools and machinery that are needed. For some instructions, it's not enough merely to list equipment and supplies. You may also have to specify such things as sizes, brands, types, and model numbers.

Structure of the instructions. Before you dive into the step-by-step discussion, identify the *tasks* in the procedure. Your instructions may have a simple series of steps that readers perform in sequence. For example, changing the oil in a car involves one task, a series of steps that must be performed in order—otherwise, you've got oil all over the driveway, a burned-up engine, or both.

However, some instructions may describe several tasks that can be performed in practically any combination or order. Operating voice mail involves numerous tasks, some of which you perform only occasionally (recording a new greeting); others, every day (playing back messages or deleting messages). If that's the case, then use headings to help readers to find these tasks quickly.

Discussion of the steps. When you discuss the individual steps (the individual actions readers take to accomplish the procedure), be aware of some issues involving writing style, format, headings, and content:

- *Imperative writing style.* In instructions, many sentences use the imperative (for example, "Press Enter" or "Calculate the square footage"). Many other sentences are phrased with the word "you" (for example, "You should check the temperature of the . . . "). Don't hesitate to use this "in-your-face" style of writing; address readers directly, get their full attention, and be straightforward about what they are supposed to do.

- *Supplemental explanation.* Some individual steps may require additional explanation. You may need to define potentially unfamiliar terms or describe how things look before, during, or after individual steps. Notice in other instructions that bold is used to visually separate the actual instruction from the supplemental explanation.

- *Special format.* When you explain the individual steps, use numbered lists for sequential steps. Use bulleted lists for steps in no necessary order (for example, nonsequential troubleshooting steps). The vertical-list format helps readers follow the procedure and visually cues them for each specific action to perform.

- *Headings.* For all but the simplest instructions, use headings. Headings enable readers to find equipment lists, background information, and troubleshooting tips. Headings guide readers to specific tasks. For example, using well-designed headings in voice-mail instructions, readers can quickly find the section on how to forward messages.

- *Graphics.* Sometimes words cannot convey enough detail about key objects and key actions. For example, just how does part A fit into part B? Make a list of the *key objects* and the *key actions* in your instructions, and identify those that readers might have trouble with. These are the ones for which you may need graphics (see Chapter 7).

RECOMMENDATION REPORTS

A recommendation report evaluates or promotes an idea—for example, the possibility of employee telecommuting. The context can vary: Management might direct you to study the feasibility of telecommuting and to make recommendations, or management might direct you to compare telecommuting products and to recommend one. The common element is a recommendation and a comparative discussion that supports that recommendation. Where you work, it may be called a recommendation report, an evaluation report, a feasibility report—or even a proposal. But the essential structure is the same for all—comparing options and recommending one.

SOME DISTINCTIONS

A recommendation report, as its name indicates, makes a recommendation about plans, products, or people. In its simplest form, it establishes certain requirements (often called criteria), compares two or more options, and recommends one. Other elements may be involved: background on the technology; descriptions of the options; explanation of how the field was narrowed; even discussion of the technical, economical, and social practicality of the idea.

Typically, the terms *proposal*, *feasibility report*, *evaluation report*, and *recommendation report* are used interchangeably. Don't expect much precision in real-world usage of these terms. Here are some distinctions:

- **Recommendation report.** Compares two or more options against each other (and against certain requirements) and then makes a recommendation.
- **Evaluation report.** Compares an idea, program, or thing against criteria or requirements as a means of determining its value. This type may recommend, but essential is the statement of the value of the idea, program, or thing.
- **Feasibility report.** Compares a project against requirements relating to its economic, technical, or social practicality (or all three), and then recommends whether the project should be initiated.
- **Proposal.** Makes a bid or seeks approval to do a project and then presents the proposer's qualifications. Its primary task is to land a contract or get approval.

Each of these types works toward an endorsement, recommendation, or value judgment; your job as the writer is to achieve that end. To write this type of report, remember that you must provide data and conclusions so that readers can decide for themselves whether your recommendations are justified.

Introduction.   As with any introduction, indicate the purpose of the recommendation report. Indicate right up front that the purpose is to recommend something for some specific situation. Indicate the audience—the intended readers of the report and any technical background they need (and if none is needed, say so). Also provide a sentence briefly listing the contents of the report.

Background on the situation.   Consider whether to discuss the situation in which this report is needed. The immediate audience may know perfectly well what the situation is, but your report may get passed around to others who don't. Background may also prove a helpful memory jogger for overly busy readers.

Requirements.   In any recommendation, there are requirements such as cost, operational features, size, and weight. Consider the example of selecting software for telecommuting: What are the specifications? Ease of use? Versions for Macintosh and PC machines? Desktop-sharing capability? Audio? In your recommendation *study*, you determine these requirements. In your recommendation *report*, you describe these requirements. Readers can then consider these requirements and decide for themselves whether they agree.

Technical background.   It may be necessary to provide some brief technical discussion. In the early days of CD-ROM products, you might have discussed 32- and 64-bit technology, triple or quadruple spin, sampling, and other related technical concepts. As with most introductory and background sections, it's best to write them later in your process. Write the heart of the recommendation report first—the comparisons, conclusions, and recommendations. (See the requirements for water purity in pharmaceutical manufacturing in Figure 5-12.)

Description.   In some recommendation reports, you may need to describe the options you are comparing. These descriptions are neutral—no comparisons or evaluations are provided. For example, in a recommendation on hybrid electric automobiles, describe each one separately in terms of its size, dimensions, range, charging requirements, and so on. (In Figure 5-12, see the description of one of the water-purification systems being compared.)

Point-by-point comparisons.   Comparisons constitute one of the three main sections of a recommendation report. For example, in a section on cost, the cost of each option is compared. Usually, it's not a simple matter of one being the cheapest and another being the most expensive. Things get blurred by special features and service plans that can be added. In these cases, help readers: Untangle the complexities for them and point to the best choice.

End each comparative section with a statement as to which option is best in terms of that comparative category. In Figure 5-13, the comparison categories are capacity, microbial prevention ability, chemical removal ability, and environmental impact.

Remember: You write these comparisons so that readers can see your logic—how you reached your conclusions. Give readers a chance to disagree with your thinking and to reach their own conclusions and recommendations.

Conclusions (summary).   The conclusions section is a repeat of the conclusions you reached in each of those individual comparative sections. For example, in the microbial prevention section, the RO system is declared high in terms of microbial water quality. This declaration is repeated along with others in the conclusions section.

In some cases, no individual option may prove to be the obvious, best choice. One option may be the cheapest; another may be the most reliable; another may be the easiest to use; still another may have far more functions and features. These are the *primary conclusions*. But how do you pick a "winner" when they conflict? If you've defined them carefully, your requirements should point to the final recommendation. Requirements enable you to state *secondary conclusions*: conclusions that resolve conflicting primary conclusions. In Figure 5-13, notice that conclusion 8 states that although the RO system has high installation costs, those costs can be depreciated, which thus offsets this particular negative.

Recommendation: Water Purification System for a Clinical Production Facility

This report investigates the appropriate water systems that are used for pharmaceutical production operations and ensure that they meet capacity requirements, quality recommendations and are in compliance with current Good Manufacturing Practices (cGMP), as defined by regulatory authorities. It will evaluate steam de-ionizer purified water systems, resin-based purified water systems, and reverse-osmosis purified water systems.

Background: Standards for Water Purity

The FDA, EMA and MHRA require that water used in the production of pharmaceutical produc[...] certain standards, depending on the class of products produced. F[...] high standard of purity[...] and creams requires a[...] solid dosage form prod[...] specifications outlined[...] water from this system[...] purified water must be[...] standard, set by the US[...] electric current passed[...] at a value measured in[...] resistivity must equal o[...] from microbial contami[...] commissioned and vali[...]

Definition of Term[...]

- *cGMP.* Current good[...] authorities.
- *EMA.* European Med[...]
- *FDA.* Food and Drug[...]
- *Gpm.* Gallons per m[...]
- *MHRA.* Medicines ar[...] Kingdom.
- *RO.* Reverse osmos[...] filters and high press[...] materials to be expe[...]
- *USP.* United States [...] United States of Am[...] yearly[7].

Steam Still De-Ionizer Systems

The still is a simple and inexpensive way for producing de-ionized water. It uses potable tap water [8] as a feed supply without expensive microbial pretreatment filtering [2] due to the fact that the distillation process prohibits microbial growth. Only a water softener is required for removing chlorine [12]. The process for producing de-ionized water begins in the boiling chamber where a heating element boils water to capture the steam. The steam leaves behind the ions and other chemical impurities while the heat destroys microbial organisms [3,5]. The steam is pumped through a series of heat exchangers condensing into pure de-ionized water [16]. Figure 1 is a typical steam still de-ionizer system.

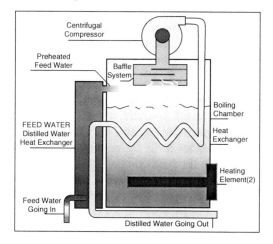

Figure 1. Vapor Compression Distiller, Vapor Compression Water Distillation Systems, Aqua Technology For the 21st Century. June 2, 2012. www.aquatechnology.net/vaporcompressiondistillers.html.

Figure 5-12 Recommendation report. Introduction, background, glossary, description of one of the water-purification systems being compared.
Source: Adapted with permission from Brian Kynast, Senior Technical Writer, Aerotek, Inc.

Recommendations.   The recommendations section simply states what has probably become obvious—which option is recommended. The example in Figure 5-13 briefly mentions which comparisons were most influential in reaching the final recommendation.

Points of Comparison

While the overall installation and operation costs are impacted by the specific needs of the organization, they all share advantages and disadvantages in regard to not only cost but quality and capacity. The comparisons are discussed below.

- *Capacity:*
  Each of the systems varies greatly in terms of output in gallons per minute. Steam stil
  condensa
  some of th
  ranges fro
  to 10 gpm

- *Microbial |*
  The steam
  the steam
  distributio
  systems c
  limits the
  through-p
  heat exch
  still a risk
  piping. Th
  provides s
  microbial |

- *Chemical |*
  The steam
  which just
  for a few r
  the thresh
  can produ
  the capaci
  increase t
  to the resi
  size, allow
  meg-ohms
  the final st
  oral solid

- *Environme*
  While all p

Table 1. Comparison of the Different Purified Water Systems

Comparative Category	Steam De-ionizer	Resin System	RO System
Capacity Output	1 Meg-ohm	18 Meg-ohm	20 Meg-ohm
Continuous Flow	No	Yes	Yes
Microbial Water Quality	Low	Medium	High
Pretreatment Expense	None	High	High
Unit Price	Low	Moderate	High
Labor Intensive	No	Yes	No
Installation Cost	Low	Moderate	High
Maintenance Cost	Low	Moderate	High
Energy & Resource Use	110V/Steam	220V	220V

Conclusions

1. The steam distillation system requires little resources and is low in installation costs.
2. The steam distillation system does not have the desired capacity, nor is it designed for a continuous flow distribution to reduce the potential for microbial contamination.
3. The resin-based system is labor intensive, given the need to continuously replenish the system with hazardous chemicals for regeneration.
4. The waste for the hazardous chemicals involved with the resin-based system is an EPA problem.
5. The reverse osmosis system provides a stable and consistent flow of water at a nominal capacity of 3 gpm.
6. With added polishing canisters, this system provides well above the USP Purified Water limit of 1 meg-ohm.
7. The RO system requires little maintenance and has no chemical regeneration requirements. Table 1 provides a summary of the differences between the water systems.
8. The RO system's higher installation costs can be depreciated over the expected life of the system and do not outweigh the benefits of this cost-effective system.
9. Although upfront installation costs are high, the reverse osmosis system is the best choice because of its reliability, capacity, moderate maintenance costs and ability to consistently achieve water quality standards.

Recommendation

While the upfront installation costs are high, I recommend the reverse osmosis system because of its reliability, capacity, moderate maintenance costs, and ability to consistently achieve water quality standards. Its higher installation costs can be depreciated over the life of the system and do not outweigh its benefits.

Sources

1. USP29 Section 1231 Water for Pharmaceutical Purposes. United States Pharmacopeia, Subsection "Purified Water", www.pharmacopeia.cn/v29240/usp29nf24s0_c1231.html.

Figure 5-13 Recommendation report. Point-by-point comparisons, summary table, conclusions, recommendation, and sources.

Sometimes, recommendations may not be so obvious either. To make recommendations, you may have to state qualifications. If you were comparing hybrid vehicles, for example, you might use conditional statements like these:

- *If your primary concern is emissions control above all other factors, choose the Insight.*
- *If your priorities are performance and safety features, choose the Civic.*
- *If you want a balance of low emissions, performance, and safety, choose the Prius.*

Sometimes, you may not be able to recommend *any* of the options whole-heartedly. Imagine doing a study in the early days of hybrid vehicles. You investigate as many of the different applications as you can, read the reviews, and get as many product demos as you can. At the end, you throw up your hands, and recommend waiting for the technology to develop.

> ### *Lotus flower power*
>
> Monarch Power, an Arizona tech company, has designed a device that unfolds 18 petals to form a 4-meter diameter flower solar collector that may be able to produce 3 kW of photovoltaic electrical power in ideal conditions.
>
> For details, see the Preface for the URL.

EXERCISES

Talk to several professional engineers about the reports they write, and ask them questions like the following:

1. Which of these types of reports do they most commonly write? Are there other types, not covered in this chapter, that they also write?

2. What are the chief purposes of the reports they write? Are the reports for internal or external consumption, for colleagues or clients?

3. How important are these reports to their business and professional careers? How much, for example, do they rely on proposals to get contracts? How often is their entire work product simply a written document?

4. Do they get editorial or production assistance in preparing these reports? Or do they get editorial help?

5. Ask your engineering interviewees about the progress reports they write. What sorts of projects require progress reports? How often do they submit progress reports for a typical project? Are the requirements stated in the contract?

BIBLIOGRAPHY

American National Standards Institute. *American National Standards for Product Safety Signs and Labels*. ANSI Z535.4-2011. New York: American National Standards Institute, 2011.

Caher, J.M. Technical Documentation and Legal Liability. *Journal of Technical Writing and Communication*, *25* (1995): 5–10.

Clement, D.E. Human Factors, Instructions, Warnings, and Product Liability. *IEEE Transactions on Professional Communication*, *30*(3) (1987): 149–156.

Dolphin, W.D. Writing Lab Reports and Scientific Reports. www.mhhe.com/biosci/genbio/mader inquiry/writing.html. Accessed March 3, 2012. (If this page is no longer available, see the Preface for the web address of the companion website for this textbook.)

Fitchett, P. and J. Haslam. *Writing Engineering Specifications*. Spon Press, 2002.

Heylar, P.S. Products Liability: Meeting Legal Standards for Adequate Instructions. *Journal of Technical Writing and Communications*, *22*(2) (1992): 125–147.

John Oriel. Guide to Specification Writing for U.S. Government Engineers. http://nawctsd.navair .navy.mil/resources/library/acqguide/spec.htm

Virginia Tech. Laboratory Reports. www.writing.eng.vt.edu/workbooks/laboratory.html. Accessed March 3, 2012. (If this page is no longer available, see the Preface for the web address of the companion website for this textbook.)

6

WRITING RESEARCH AND DESIGN REPORTS

Reports are perhaps the most common documents that you will write both as engineering students and as engineers. Consequently, your success—both in school and in the workplace—will partly depend on your ability to produce effective reports.

Susan Stevenson and Steve Whitmore, *Strategies for Engineering Communication* (New York: John Wiley & Sons, 2002).

This report, by its very length, defends itself against the risk of being read.

Winston Churchill.

This chapter covers two important engineering report types: the information type, called here the *engineering report*, and the *engineering design report*. This chapter ends with format and style issues relevant to both types of reports.

ENGINEERING RESEARCH REPORTS

Engineers often get involved in projects that include writing reports. Engineering reports have specifications just like any other kind of project. *Specifications* for reports involve layout, organization and content, format of headings and lists, design of the graphics, and so on. In fact, the American National Standards Institute (ANSI) has published specifications for engineering reports entitled *Scientific and Technical Reports: Organization, Preparation, and Production.*

The advantage of a required structure and format for reports is that you or anyone else can expect them to be designed in a familiar way—you know what to look for and where to look for it. Reports are usually read in a hurry—people are in a hurry to get to the information they need, the key facts, the conclusions, and other essentials. A standard report format is like a familiar neighborhood.

When you analyze the design of an engineering report, notice how repetitive some sections are. That's because people don't read reports straight through: They may start with the executive summary, skip around and probably not read every page. Your challenge is to design reports so that these readers encounter your key facts and conclusions, no matter how much of the report they read or in what order they read it.

> **Solar paint your house**
>
> Researchers at Notre Dame's Center for Nano Science and Technology are developing a paste that, when brushed onto a transparent conducting material and exposed to light, produces electricity!
>
> For details, see the Preface for the URL.

The standard components of the typical engineering report are as follows:

Transmittal letter

Covers and label

Table of contents

List of figures

Executive summary

Introduction

Body of the report

Conclusions

Appendixes (including references)

The following sections guide you through each of these components, pointing out the key features. As you read and use these guidelines, remember that these are *guidelines*, not commandments. Different professions and organizations have their own guidelines for reports—adapt your practice to those as well the ones presented here.

Note The example used in this chapter is a *background* report. However, many resources point to the research report as *the* engineering report. Discussion and examples of the research report can be found in Chapter 5. Additional examples of both types are available at the companion website for this book. See the Preface for the web address.

LETTER OF TRANSMITTAL

The transmittal letter is a cover letter. An example is shown in Figure 6-1. It is either attached to the outside of the report or bound within the report. It is a communication from you—the report writer—to the recipient, the person who requested the report and who may even be paying you for your expert consultation. Essentially, it says, "Okay, here's the report that we agreed I'd complete by such-and-such a date.

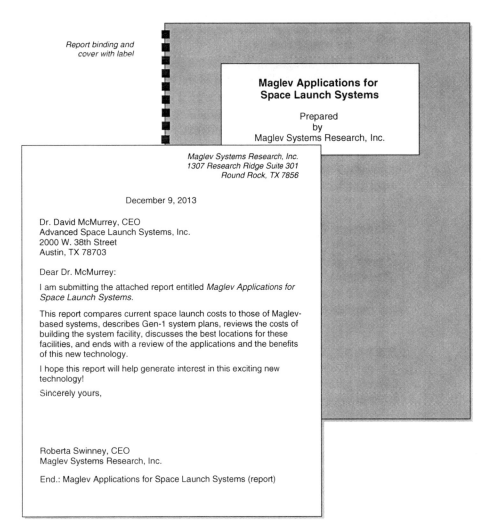

Figure 6-1 Transmittal letter and report front cover. Often an informal transmittal letter is attached to the front of the report, whereas a more formal one is bound in the report as part of its permanent history.

Continuing, the transmittal letter says essentially: "Briefly, it contains this and that, but does not cover this or that. Let me know if it meets your needs." The transmittal letter explains the context—the events that brought about the report. It contains information about the report that does not belong in the report.

In Figure 6-1, notice the standard business-letter format. If you write an internal report, use the memorandum format instead; in either case, the contents and organization are the same:

- **First paragraph.** Cites the name of the report, putting it in italics. Either in the first paragraph or in the following, you can also mention the reason for the report and the date it was assigned.

- **Second paragraph.** Focuses on the purpose of the report and briefly overviews the report contents.
- **Third paragraph.** Acknowledges any funding or help by other people and also mentions any limitations of the report.
- **Final paragraph.** Closes with a gesture of good will, expressing hope that the reader finds the report satisfactory. It also encourages the reader to get in touch with any questions, comments, or concerns, and gives contact information.

As with any other element in an engineering report, you may need to modify these contents. For example, you might want to add another paragraph, listing questions you want readers to consider as they review the report.

REPORT COVER AND LABEL

If your report is over ten pages, bind it in some way and create a label for the cover. (Remember that the preferred alternative may be to create a portable document using an application such as Adobe Acrobat. See "Generating Portable Documents" later in the chapter.)

REPORT COVERS

Covers give reports a solid, professional look as well as protection. You can choose from many types of covers. Here are some recommendations:

- Totally unacceptable are the clear (or colored) plastic slip cases with the plastic sleeve on the left edge. These are like something out of freshman English; plus they are aggravating to use.
- Marginally acceptable are the covers for which you punch holes in the pages, load the pages, and bend down the brads. Leave an extra half-inch margin on the left edge so that readers don't have to pry the pages apart. Annoyingly, readers must grab available objects or use various body parts to keep the pages weighted down.
- By far the best covers are those that allow reports to lie open. This type uses a plastic spiral for the binding and thick, card-stock paper for the covers. Check with your local copy shop for these types of bindings; they are inexpensive and add to the professionalism of your work.
- Less preferable are looseleaf notebooks, or ring binders. These are too bulky for short reports, and the page holes tend to tear. True, the ring binder makes changing pages easy; if that's how your report will be used, it's a good choice.
- At the "high end" are the overly fancy covers with their leatherette look and gold-colored trim. Avoid them—keep it plain, simple, and functional.

REPORT LABELS

Be sure to devise a label for the cover of your report. It's a step that some report writers forget. Without a label, a report is anonymous; it gets ignored.

The best way to create a label is to use your word-processing software to design one on a standard page with a graphic box around the label information. Print it out, then go to a copy shop and have it photocopied directly onto the report cover.

Not much goes on the label: the report title, your name, your organization's name, perhaps a report tracking number, and a date. Check with your organization for requirements. (An example of a report label is shown in Figure 6-1.)

ABSTRACT AND EXECUTIVE SUMMARY

Most engineering reports contain at least one abstract—sometimes two, in which case the abstracts play different roles. Abstracts summarize the contents of a report, but the different types do so in different ways:

- **Descriptive abstract.** Provides an overview of the purpose and contents of the report. In some report designs, the descriptive abstract is placed at the bottom of the title page.
- **Executive summary.** Summarizes the key facts and conclusions contained in the report. (See the example shown in Figure 6-2.) Typically, executive summaries are one-tenth to one-twentieth the length of reports 10 to 50 pages long. For longer reports, ones over 50 pages, the executive summary should not go over three typewritten pages. The point of the executive summary is to provide a summary of the report—something that can be read quickly.

Note   Check within your organization to determine which types of summaries are required and what they are called.

If the executive summary, introduction, and transmittal letter strike you as repetitive, remember that readers don't necessarily start at the beginning of a report and read page by page to the end. They skip around. For these reasons, reports are designed with some duplication so that readers will be sure to see the important information no matter where they dip into the report.

TABLE OF CONTENTS

You're familiar with tables of contents (TOC) but may never have stopped to look at their design. The TOC shows readers what topics are covered in the report, how those topics are discussed (the subtopics), and on which page numbers those sections and subsections start.

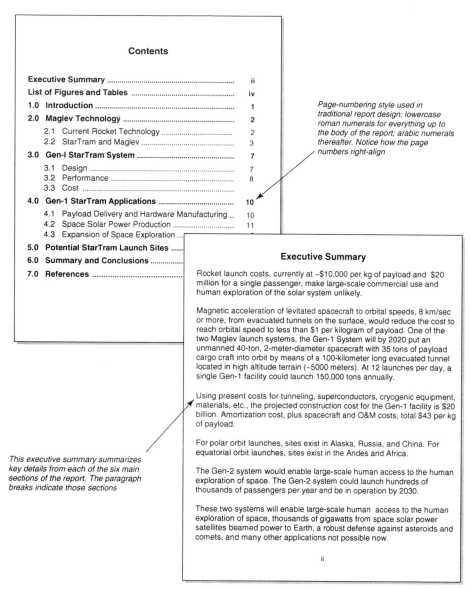

Figure 6-2   Table of contents and executive summary from an engineering report. (Some reports include a descriptive abstract at the top or bottom of the title page.)

In creating a TOC, you have a number of design decisions:

- **How many levels of headings to include.** In longer reports, don't include all of the lower-level headings; otherwise, the TOC will be too long and unwieldy.

The TOC should provide an at-a-glance way of finding information in the report quickly.

- **Indentation, spacing, and capitalization.** In Figure 6-2, notice that there are two levels of headings and that the first level is bolded but not the second. Also notice the right alignment of the page numbers. Notice also that the main chapters or sections use initial caps on each main word; lower-level sections use initial caps on the first word only.
- **Vertical spacing.** Notice that the first-level sections have extra space above and below, which increases readability.

One final note: Make sure the words in the TOC are the same as they are in the text. As you write and revise, you might change some of the headings—don't forget to change the TOC accordingly. (Automatically generating your TOC will take care of this problem.)

Note   See the chapter in the website companion corresponding to this one for steps on automatically generating TOCs and lists of figures and tables for your reports. For the web address, see the Preface.

LIST OF FIGURES AND TABLES

The list of figures has many of the same design considerations as the table of contents. Readers use the list of figures to find the illustrations, diagrams, tables, and charts in your report.

Complications arise when you have both tables and figures. Strictly speaking, *figures* are illustrations, drawings, photographs, graphs, and charts. *Tables* are rows and columns of words and numbers; they are not considered figures.

For longer reports that contain dozens of figures and tables each, create separate lists of figures and tables. Put them together on the same page if they fit, as shown in Figure 6-3.

> **Solar-powered boat**
>
> A Canadian eco-enthusiast has designed a solar-powered boat that can accommodate six passengers. Its 8 x 6 panel of photovoltaic cells produces 140W.
>
> For details, see the Preface for the URL.

INTRODUCTION

An essential element of any report is its introduction—make sure you are clear on its real purpose and contents. The introduction prepares readers to read the main body of

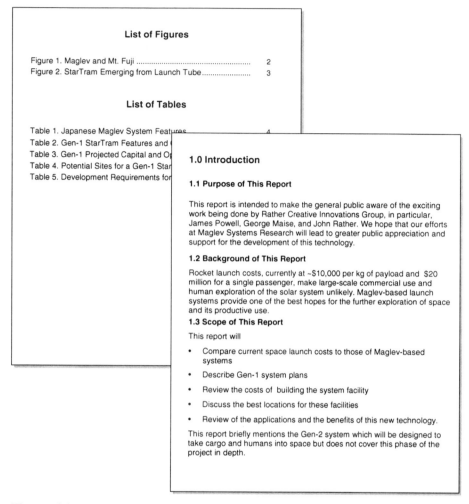

1.0 Introduction

1.1 Purpose of This Report

This report is intended to make the general public aware of the exciting work being done by Rather Creative Innovations Group, in particular, James Powell, George Maise, and John Rather. We hope that our efforts at Maglev Systems Research will lead to greater public appreciation and support for the development of this technology.

1.2 Background of This Report

Rocket launch costs, currently at ~$10,000 per kg of payload and $20 million for a single passenger, make large-scale commercial use and human exploration of the solar system unlikely. Maglev-based launch systems provide one of the best hopes for the further exploration of space and its productive use.

1.3 Scope of This Report

This report will

- Compare current space launch costs to those of Maglev-based systems
- Describe Gen-1 system plans
- Review the costs of building the system facility
- Discuss the best locations for these facilities
- Review of the applications and the benefits of this new technology.

This report briefly mentions the Gen-2 system which will be designed to take cargo and humans into space but does not cover this phase of the project in depth.

Figure 6-3 List of figures and introduction for an engineering report. (The introduction comes *after* the list of figures.)

the report. It does not dive into the subject, although it may provide a bit of theoretical or historical background. Instead, introductions indicate or discuss the following (but not necessarily in this order):

- Specific purpose and topic of the report (indicated somewhere in the first paragraph).
- Intended audience of the report—knowledge or experience that readers need in order to understand the report.
- Situation that brought about the need for the report.

- Scope of the report—topics covered as well as topics not covered (specifically, ones that some readers might expect).
- Background (such as concepts, definitions, history, statistics)—just enough to get readers interested; just enough to enable them to understand the context.

Review the introduction in Figure 6-3 to see how these elements are handled.

The introduction is often mistakenly considered to be synonymous with background information. As the preceding list shows, background is only a minimal part of an introduction. Remember: The introduction prepares readers to read the report; it "introduces" them to the report. If the background gets out of hand and runs on for too many paragraphs, move it to a section of its own, either just after the introduction or into an appendix. For a typical 20-page report, for example, the introduction should be no more than two pages—and the background within the introduction no more than a third of the introduction.

BODY OF THE REPORT

The body of the report is of course the main text of the report, the sections between the introduction and conclusion. Figure 6-4 shows several sample pages. Notice the bracketed 2 in Figure 6-4; it corresponds to the sources list in Figure 6-5.

CONCLUSIONS

For most reports, you'll need to include a final section. When you plan the final section of an engineering report, think about the functions it can perform in relation to the rest of the report:

- **Conclude.** Draw (or repeat) logical conclusions from the discussion that has preceded; make inferences on the data that has been presented.
- **Summarize.** Review the key facts and ideas from the preceding material. Summaries present nothing new—they leave readers with a perspective on what has been discussed, the perspective that you want them to have.
- **Generalize.** Move away from the specific topic of the report to a general discussion of such implications, applications, and future developments—but only in general terms.

Your final section can do any combination of these, depending on your sense of what your audience and report need. The example conclusion in Figure 6-6 summarizes the key conclusion contained in the report, speculates about housing trends, and takes a brief look at recent developments.

The length of the conclusion can be anything from a 100-word paragraph to a 5- or 6-page section. For the typical 10- to 20-page report, the final section is 1 to 2 pages,

2.2 StarTram and Maglev

The first StarTram system, Gen-1, is a high G cargo launch system. After reaching orbital speed, the vehicle leaves the acceleration tunnel at a high altitude, but still at ground level. The vehicle then coasts up to orbit, experiencing strong but manageable aerodynamic heating and deceleration forces. Because of the low energy cost per kilogram, large amounts of protective coatings and coolants for the cargo craft do not significantly increase launch cost.

Figure 2. StarTram Emerging from Laur

Full evelopment of the Gen-1 will require e studies in the following areas:

- Blunt nose geometry, with an effective
- Different launch altitudes (4000, 6000,
- Different launch angles (10 or 15 degr
- Three different approaches to compen ascent through the atmosphere
- Gen 1 specifications: 40-ton cargo cra long (currently)

The $20 billion construction cost is small compared with other government programs. The NASA Constellation program for the return to the Moon is ~100 billion dollars. Gen-1 is approximately 2 weeks of the US defense budget [2]. The economic benefits will greatly outweigh its cost.

Table 2. Gen-1 StarTram Features and Capabilities

- Gen-1 Maglev levitation and LSM propulsion based on Powell-Danby inventions and the operating Japanese Maglev system
- Mature, reliable, low-cost Nb-Ti superconductor and cryogenic system technology used
- Electric launch energy generated by a conventional power plant and stored for acceleration in Superconducting Magnetic Energy Storage (SMES) loops
- Exit from an evacuated acceleration tunnel closed to the atmosphere by a mechanical shutter and outer thin plastic film
- Gen-1 cargo craft reference design:
  - 2 meters outer dimensions, 13 meters long, 8 km/sec velocity
  - 40 metric ton total weight; 35 ton payload
  - 10 launches daily; 128,000 tons of payload per year
  - 30 G acceleration

5.0 Gen-1 StarTram Applications

5.1 Payload Delivery and Hardware Manufacturing

The amounts of payload using Maglev launch are much greater than present rocket systems can provide. Besides reducing the launch cost per kilogram to less than 1/100th of present costs, Maglev launch will also greatly reduce the manufacturing cost of space hardware. The very high launch cost of present rocket systems forces the hardware to be very reliable, ultra-lightweight, and operate close to its limits, "one of a kind" systems with extremely expensive engineering and quality control. With very low cost and unlimited launch capability of Maglev launch, space hardware could be much cheaper using existing, heavier components instead of new one-of-a-kind. Hardware would not have to operate close to failure limits.

Figure 6-4   Pages from the body of an engineering report. Note the use of headings, tables, and the citation (the ''[2]'') of a borrowed information source.

but such ratios should never be applied without considering the report. Watch out for conclusions that get out of hand and become too long. Readers expect a sense of closure, a feeling that the report is ending. When the final section becomes too long, consider doing one of the following: Move the discussion back into the body of the report; shorten and generalize the discussion in order to keep it in the conclusion; or find some other way to end the report.

Documenting your information sources is all about establishing, maintaining, and protecting your credibility in the profession. You must cite (''document'') the sources

6.0 Summary and Conclusions

This report has described a new way to launch large payloads into orbit and beyond. The StarTram Maglev launch system uses magnetically levitated vehicle technology similar to that now operating for passenger transport in Japan.

Two Maglev launch systems have been planned: a near-term Gen-1 system that will launch a 40-ton cargo craft to high altitude (4000 meters) at ~8 km/sec from an evacuated acceleration tunnel. A Gen-1 facility can launch 100,000 tons or more of payload annually, at a unit cost of less than $50 per kilogram. The long-term Gen-2 system that will launch humans and cargo into space, using the same technology, was briefly mentioned.

Maglev launch technology will have several important applications:

- Solar power satellites continuously beaming hundreds of gigawatts of electric power to Earth from GEO orbit—at far less cost than fossil fuel and nuclear power plants.

- Large manned bases on the Moon and colonies on Mars.

- Highly detailed exploration of the outer solar system using low-cost robotic probes, with samples returned to Earth for analysis.

- Strong defense against asteroids and long-period comets threatening Earth.

ferences

High Speed Transport by Magnetically
-WA/RR-5. *ASME Meeting*, NY, NY.
(1967).

iagua, J. "StarTram: A New Concept for
Transport Using Ultra High Velocity
-01-S.6.04, 52nd International
use, France, Oct. 1-5 (2001).

iagua, J. "StarTram: A Maglev System
Cargo to LEO, GEO, and the Moon",
th International Congress, Bremen,

iagua, J. "StarTram: An Ultra Low Cost
ge Scale Exploration of the Solar
nd Applications International Forum
NM, February 12-16 (2006).

a, J., and Jordan, J., "StarTram – An
tically Launch Payloads at Ultra Low
2.7; *57th International Astronautical Congress*, Valencia, Spain, October (2006).

[6] Carlson, H.W. "Simplified Sonic Boom Prediction", *NASA TP-1122*, March (1978).

[7] Ishmael, S.D. "What is the X-30?", in *Proceedings of the First flight 30th Anniversary Celebration*, NASA Hugh L. Dryden Flight Test Research Facility, Edwards, California, January (1991).

[8] Powell, J., Maise, G., Paniagua, J., and Rather, J. "Magnetically Inflated Cable (MIC) System for Large ScaleSpace Structures", *NIAC Phase 1 Report*, May 1, 2006, NIAC Subaward No. 07605-003-046. Also, "MIC – A Self Deploying Magnetically Inflated Cable System", *Acta Astronautica*, 48, No 5-12, p 331-352.(2001).

Figure 6-5   Conclusion and references page. Notice that the summary and conclusions page summarizes the chief findings of the report, speculates on those findings, and then glances at more recent developments. The references page uses the IEEE system of documenting borrowed information. Also see the bracketed citations in Figure 6-4. (See Chapter 11 for details on the IEEE system.)

of borrowed information regardless of the shape or form in which you present that borrowed information. Whether you directly quote it, paraphrase it, or summarize it—it's still borrowed information. Whether it comes from a book, article, a diagram, a table, a web page, a product brochure, an expert whom you interview in person—it's still borrowed information.

See Chapter 11 for details on how to cite the sources of your borrowed information using the IEEE documentation system.

Executive Summary

We began by dissecting an automat
green design and eliminate the need
connections.

For our first design review, we exam
be designed to require less power, d
but decided to start over with an ent

The final design, the thermal ice mal
discovered during our user testing: n
little user interaction.

The most novel part of the design is
its low weight, low price, and high co
staggered wells so that the water flo
automatically fills each well to the sa
overflow.) Each well is a triangular p
side so that each "cube" has the ma

When the ice's weight overcomes th
separates and falls into a bin with no
heating coils or motors. According to
water as cold as 3 degrees Celsius.

Outside of the tray, we designed an
axis-support to protect the flavor of t
pleasing appearance.

It is cheap to make and assemble, s
fifty dollars, and the tray alone for un
Considering an automatic ice maker
purchase and installation costs and
have a heating coil heating their free
is, in fact, a marketable invention.

Thermal Release Ice Maker

Engineering Design II – Carnegie Mellon University

Erica Dorfman, Eleanor McDaniel, Andrew Moore, J. Nicholas Smarto

Figure 1. Final Design Model

Table of Contents

Figure 6-6   Table of contents and executive summary of a design report.

ENGINEERING DESIGN REPORTS

An engineering design report is written to introduce and document engineering design projects. Audiences include other engineers interested in the functions and effectiveness of the design and management interested in how the design can be applied and commercialized. Engineering students, for example, have written about the design of a temperature measurement and display system using a microcontroller and the design of a device to simulate flow through a ribbed cooling passage. Engineering students at Carnegie Mellon have designed a thermal-release ice maker.

Net-zero energy home

CHIP House (Compact Hyper-Insulated Prototype) was developed by students and a partnership between Caltech and SCI-Arc. As a net-zero energy home, it requires no external energy source.

For details, see the Preface for the URL.

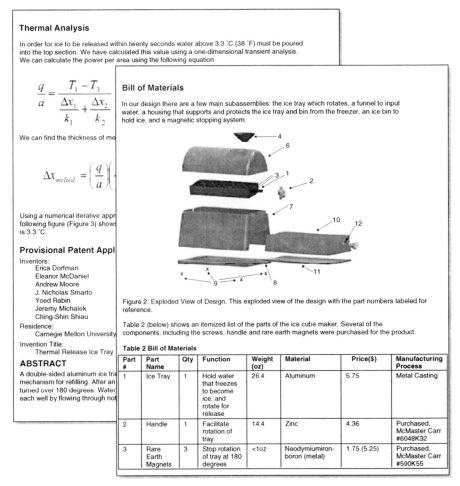

Thermal Analysis

In order for ice to be released within twenty seconds water above 3.3 °C (38 °F) must be poured into the top section. We have calculated this value using a one-dimensional transient analysis. We can calculate the power per area using the following equation

$$\frac{q}{a} = \frac{T_1 - T_3}{\frac{\Delta x_1}{k_1} + \frac{\Delta x_2}{k_2}}$$

We can find the thickness of me

$$\Delta x_{melted} = \left(\frac{q}{a}\right)$$

Using a numerical iterative appr
following figure (Figure 3) show
is 3.3 °C.

Provisional Patent Appl

Inventors:
  Erica Dorfman
  Eleanor McDaniel
  Andrew Moore
  J. Nicholas Smarto
  Yoed Rabin
  Jeremy Michalek
  Ching-Shin Shiau
Residence:
  Carnegie Mellon University

Invention Title:
  Thermal Release Ice Tray

ABSTRACT

A double-sided aluminum ice tra
mechanism for refilling. After an
turned over 180 degrees. Water
each well by flowing through not

Bill of Materials

In our design there are a few main subassemblies: the ice tray which rotates, a funnel to input water, a housing that supports and protects the ice tray and bin from the freezer, an ice bin to hold ice, and a magnetic stopping system.

Figure 2: Exploded View of Design. This exploded view of the design with the part numbers labeled for reference.

Table 2 (below) shows an itemized list of the parts of the ice cube maker. Several of the components, including the screws, handle and rare earth magnets were purchased for the product.

Table 2 Bill of Materials

Part #	Part Name	Qty	Function	Weight (oz)	Material	Price($)	Manufacturing Process
1	Ice Tray	1	Hold water that freezes to become ice, and rotate for release	26.4	Aluminum	5.75	Metal Casting
2	Handle	1	Facilitate rotation of tray	14.4	Zinc	4.36	Purchased, McMaster Carr #6048K32
3	Rare Earth Magnets	3	Stop rotation of tray at 180 degrees	<1oz	Neodymiumiron-boron (metal)	1.75 (5.25)	Purchased, McMaster Carr #590K55

Figure 6-7   Additional pages of a design report. Many thanks to the authors of this report for permission to adapt it.

The design report has many of the same parts as the engineering report presented elsewhere in this chapter. The key differences are of course the purpose of the design report and special sections that are peculiar to this type of report. You can get an immediate sense of those sections by looking at Figure 6-6, which shows the table of contents of the thermal-release ice maker. Figure 6-7 shows pages from the body of the report.

GENERAL REPORT DESIGN AND FORMAT

Characteristics such as page numbering, appendixes, cross-references, and documentation apply to either type of report discussed in this chapter.

PAGE NUMBERING

Here are some standard guidelines for numbering pages in a report:

- All pages in the report (within but excluding the front and back covers) are numbered; however, on some pages, the numbers are not displayed.
- In the contemporary design, all pages throughout the document use arabic numerals; in the traditional design, all pages *before* the introduction (first page of the body of the report) use lowercase roman numerals.
- On special pages, such as the title page and page one of the introduction, page numbers are not displayed.
- Page numbers can be placed in one of several areas on the page. Usually, the easiest choice is to place page numbers at the bottom center of the page (remember to hide them on special pages).
- If you place page numbers at the top of the page, you must hide them on chapter or section openers where a heading or title is at the top of the page.
- Longer reports often use the page-numbering style known as folio-by-chapter or double-enumeration (for example, pages in Chapter 2 would be numbered 2–1, 2–2, 2–3, and so on). This style eases the process of adding and deleting pages.

Note   See the chapter in the website companion corresponding to this one for steps on automating page numbers in your reports. For the web address, see the Preface.

CROSS-REFERENCES

You may need to point readers to closely related information within your report, or to other books and reports that have useful information. These are called *cross-references*. For example, they can point readers from the discussion of a mechanism to an illustration of it. They can point readers to an appendix where background on a topic is given (background that just does not fit in the text). And they can point readers outside your report to other information—to articles, reports, and books that contain information related to yours. When you create cross-references, follow these guidelines:

- If you refer to another section of your report, put the heading or section title in quotation marks.
- If you refer to an article in a journal or encyclopedia, put quotation marks around the article title.
- If you refer to the title of a journal, book, or report, italicize that title.
- When you create cross-references, help readers understand why they should go to that information. Indicate the topic of the cross-referenced information (don't assume the title indicates it fully), and suggest why readers might want to follow the cross-reference.

- Cite exact titles or supply page numbers if doing so helps readers. In a short report, say, one under ten pages, citing page numbers is not necessary (although word-processing software makes automating cross-references easy). If you supply the page number, then you can cite the subject matter of the section—not the exact title—in case you change the wording of headings.

Note See the chapter in the website companion corresponding to this one for steps on creating automated cross-references in your reports. For the web address, see the Preface.

APPENDIXES

Appendixes are those extra sections following the conclusion. What do you put in appendixes? Anything that does not comfortably fit in the main part of the report but cannot be left out of the report altogether. The appendix is commonly used for the following:

- Large tables of data
- Big chunks of sample code
- Fold-out maps
- Background too basic or too advanced for the body of the report
- Large illustrations that do not fit in the body of the report.

In other words, anything too large for the body of the report or too distracting and interruptive to the flow of the report is a good candidate for an appendix.

DOCUMENTATION

Documentation is the system by which you indicate the sources of the information you borrow in order to write a report. Many engineers use the system created by the Institute of Electrical and Electronics Engineers (IEEE), examples of which are shown in the figures throughout this chapter. Other engineering documentation systems vary only slightly from the IEEE system. (See Chapter 11 for details.)

GENERATING PORTABLE DOCUMENT FILES

When you develop a report, most circumstances require that you convert it to a portable document file (PDF). Consider your Word, WordPerfect, or Open Office document as the source file; send the PDF version to your colleagues and clients.

While the recipients of your PDF report can download a PDF reader, to generate the PDF you must have access to the full license of a PDF application. Adobe Acrobat is the most popular instance of PDF software. However, there are many freeware and shareware PDF applications available on the Internet.

When you generate a PDF, make sure that the resulting PDF has the following features:

- Table of contents in the side panel with links to the corresponding chapters, sections, and subsections of the report.
- Within the body of the report, cross-references to other sections of the report that are formatted as hypertext links.
- Cross-references to external Internet web pages also formatted as hypertext links.

> ### *Solar-powered airship*
>
> Solar Ship, a Canadian company, has designed a hybrid airship whose top surface is almost completely covered with solar cells.
>
> For details, see the Preface for the URL.

These matters are not always taken care of by your PDF application. Figure 6-8 illustrates the well-formatted PDF.

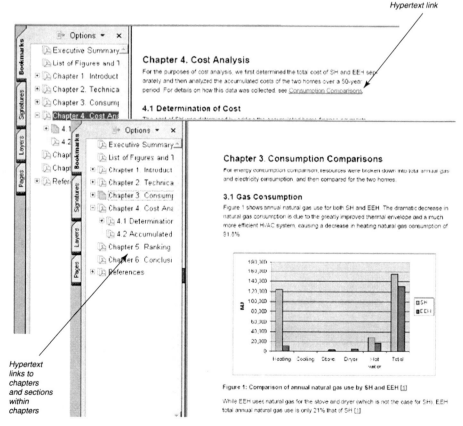

Figure 6-8 PDF version of an engineering report. The underlined text is a hypertext link to another chapter of the report. In the printed version, it would be a page number.

> **Note** See the chapter in the website companion corresponding to this one for steps on generating a PDF of your reports. For the web address, see the Preface.

USING CMS AND OTHER APPLICATIONS FOR TEAM REPORTS

A content management system (CMS) is an online application that enables people to collaborate on projects. Some are referred to as *wikis*, the first of which was created by Ward Cunningham created the first wiki in 1994, naming it after the WikiWiki shuttle bus at the Honolulu International Airport. Since then, Wikipedia, powered by MediaWiki, has become the most famous instance of group website technology—specifically, Wikipedia. Group website software, of which there are numerous varieties, can be installed on most Internet servers.

Group websites can facilitate team projects—particularly team document creation. Imagine that you are on a team of four doing an engineering project, for which there also must be a final report. A group website can facilitate your team efforts in the following ways:

- Provide an online place to upload project and report files, accessible to all team members no matter where those team members are located.
- Provide a simple interface for creating and editing documents, eliminating the need to know XHTML.
- Facilitate the team-editing of those files, particularly the report files.
- Send out email to team members when a page in the group website has been changed.
- Maintain a running narrative—problems, decisions, changes—about both the project and its report.
- Enable document control so that changes can be made by only one individual team member at a time.
- Facilitate the process of reaching consensus among team members about project issues.

Free group-website facilities, at least at the date of this publication, are also available. One such is called Google Sites, offered by Google.com. Most of the functions previously mentioned are here. Figure 6-9 shows an engineering report set up at Google Sites.

> **Note** See the chapter in the website companion corresponding to this one for steps on writing a report using collaborative writing application. For the web address, see the Preface.

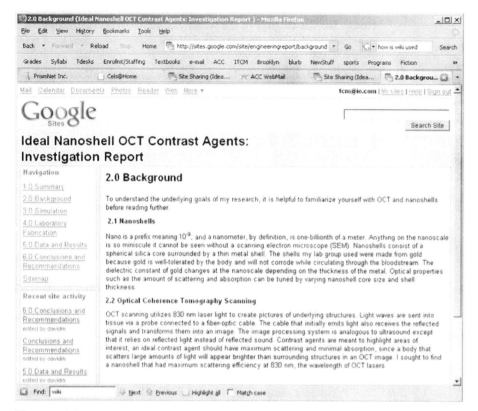

Figure 6-9 An engineering report created at Google Sites. Each team member can "own" a different section of the report, or a project coordinator can release sections for editing to individual team members. (The complete website is located at http://sites.google.com/site/engineeringreport/background, although signup is required.)

EXERCISES

Look at some examples of technical reports for the following characteristics:

1. How does the format of these engineering reports compare with the format shown in this chapter or with that specified by the American National Standards Institute's *Scientific and Technical Reports: Organization, Preparation, and Production*?

2. What are the common audiences for the reports? Are they fellow engineers or nonspecialists?

3. Typically, what purposes do the reports have? What functions do they perform for the engineering firm?

4. How are the graphics that are present in the reports created—by graphics specialists or by the engineers themselves?

5. How much are the reports a product of team writing—a group of engineers working on the project together?

6. How much library research is typically required to produce the reports? How much information for the reports comes from print and nonprint sources?

7. What process do engineering firms use in the production of reports? Do they use technical writers, graphics specialists, document designers, and editors; or is the production of reports mostly the responsibility of the engineers and clerical staff?

BIBLIOGRAPHY

If any of the following web addresses are no longer available, see the Preface for the web address of the companion website for this textbook.

American National Standards Institute. *Scientific and Technical Reports: Organization, Preparation, and Production*. ANSI Standard No. Z39.18–1987. American National Standards Institute: Washington, DC, 1987.

Beer, David. *Writing and Speaking in the Technology Professions: A Practical Guide*. 2nd Ed. New York: IEEE Press, 2003.

Penn State University, Leonhard Center. *Writing Guidelines for Engineering and Science Students*. www.writing.engr.psu.edu. Accessed July 3, 2012.

Rensselaer Polytechnic Institute, Center for Communication Practices. *ECSE Final Report*. www.ccp.rpi.edu/resources. Accessed July 3, 2012.

University of Toronto—Engineering Comunication Centre. *On-Line Handbook*. www.engineering.utoronto.ca/Directory/students/ecp/tutorial.htm. Accessed July 3, 2012.

Virginia Tech. Resources for Teaching Engineering Communication and Related Professional Skills. www.vtecc.eng.vt.edu/engcomresources.htm. Accessed July 3, 2012.

7

CONSTRUCTING ENGINEERING TABLES AND GRAPHICS

Too often writers overlook the importance of including graphics in their reports and papers. Correctly done, graphics (or visuals) not only are informative, but they also draw the readers' attention to information writers choose to highlight. . . . And if one definite trend is emerging in writing about high-tech subjects, it is an increasing reliance on visual communication.

Charles Sides, *How to Write and Present Technical Information*, 3rd ed. (Phoenix: Oryx Press, 1999), p. 48.

When you write engineering documents, you're likely to need tables, illustrations, diagrams, charts, graphs, drawings, and schematics. Nontextual material like this helps present your information more effectively and gives a polished, professional look to your work. With the increasing power and ease of use of graphics software applications, you don't need to be a graphics professional to create or adapt graphics for your engineering documents.

TABLES

You've probably constructed tables using word-processing applications such as Open-Office Write, WordPerfect, or Word. This section provides some ideas for increasing your productivity with tables and for fine-tuning the design of tables. (See Figure 7-1 for table terminology.)

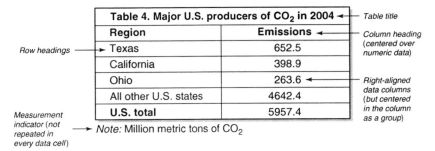

Figure 7-1 shows table terminology with these labels: Table title, Column heading (centered over numeric data), Row headings, Right-aligned data columns (but centered in the column as a group), Measurement indicator (not repeated in every data cell).

Table 4. Major U.S. producers of CO$_2$ in 2004	
Region	**Emissions**
Texas	652.5
California	398.9
Ohio	263.6
All other U.S. states	4642.4
U.S. total	**5957.4**

Note: Million metric tons of CO$_2$

Figure 7-1 Table terminology. You might prefer a table design with fewer grid lines. Check your word-processing software; it provides many different design options for tables.

Note   See the corresponding chapter in the website companion for techniques in creating tables and dynamically incorporating them into your reports. For the web address, see the Preface.

CONVERT TEXT TO TABLES

In most word-processing software, you can convert a column of text to a table (see Figure 7-2). Just make sure that you have a repeating set of elements: for example, a set of four repeating elements to create a four-column table.

USE TABLES FOR TWO-COLUMN LISTS

You have probably seen two-column lists and perhaps even created some by using

> ### *Robot to Clean Up Oil Spills?*
>
> An MIT engineer has developed Seaswarm, an unmanned system using nanotechnology and solar power to clean up oil spills—far better than the 800 skimmers that collected only 3 percent of the Deepwater Horizon leak.
>
> For details, see the Preface for the URL.

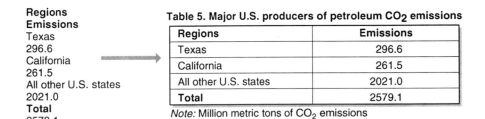

Regions
Emissions
Texas
296.6
California
261.5
All other U.S. states
2021.0
Total
2579.1

Table 5. Major U.S. producers of petroleum CO$_2$ emissions	
Regions	**Emissions**
Texas	296.6
California	261.5
All other U.S. states	2021.0
Total	**2579.1**

Note: Million metric tons of CO$_2$ emissions

Figure 7-2 Converting text to tables. Notice that the text column is arranged in groups of two. (The table title is added afterwards.)

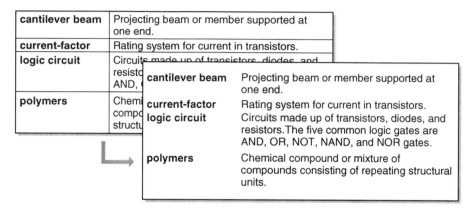

cantilever beam	Projecting beam or member supported at one end.
current-factor	Rating system for current in transistors.
logic circuit	Circuits made up of transistors, diodes, and resistors. The five common logic gates are AND, OR, NOT, NAND, and NOR gates.
polymers	Chemical compound or mixture of compounds consisting of repeating structural units.

Figure 7-3   Two-column lists—an easier way. The version on the right is still the same table; its grid lines are turned off.

tabs. Bad idea: When you add or delete words, the formatting falls apart. Instead, use a table in which you turn the grid lines off (see Figure 7-3).

IMPORT SPREADSHEET DATA TO CREATE TABLES

Many of your tables may come from data in spreadsheet applications such as Excel or Open Office Cal. There's no sense in retyping all that data—copy or import it. In most spreadsheet applications, copying is easy: Just select cells you want, copy them, and then paste them into your document (see Figure 7-4). In most applications, the pasted data cells will be formatted as a table; all you have to do is fine-tune the formatting.

Region	CO2 emission: Coal
California	6.5
Ohio	131.6
Pennsylvania	139.4
Indiana	152.6
Texas	153.4
All other U.S. states	1571.7
Total	2154.6

Table formatted in word-processing document

Original spreadsheet data

Figure 7-4   Using spreadsheets for tables. After pasting the spreadsheet data into a word-processing document, format it, as shown here.

CONVERT PARAGRAPHS TO TABLES

Study your drafts for opportunities in which plain textual discussion in paragraph format can be reworked as tables. As you can see in Figure 7-5, the same kinds of things are said about two items. Situations like these are excellent opportunities for re-presentation as tables.

FORMAT TABLES

Whichever technique you use for tables, keep these design considerations in mind:

- Include a heading at the top of each column to identify the contents of the column.
- If necessary, include row headings in the farthest left column to identify the contents of the rows.

In a comparison of Ford conventional vehicles and hybrid electric vehicles (HEV), the HEV proved to have a greater range (450-550 miles) than did the conventional vehicle (350 miles). And, as might be expected, these numbers were the same for gasoline range. In terms of fuel economy, the HEV was 30-50% better than the conventional vehicle. This, in turn, meant less frequent fill-ups for the HEV. Burning less gasoline causes the HEV to be 95% cleaner—far friendlier to the environment. And finally, this study found that the HEV performed more like a V-6 (more powerfully) than the conventional vehicle, whose performance was considered more like that of a 4-cylinder engine.

Table 1 shows the results of a comparison of conventional and hybrid electric vehicles done by Ford in 2002:

Table 1. Conventional-HEV Vehicle Comparisons		
	Conventional	Hybrid Electric
Total Range	350 miles	450-550 miles
Gasoline Range	350 miles	450-550 miles
Fuel Economy	Base	30-50% over base
Refueling	Fill-up	Fill-up (less often)
Environmental Friendliness	Base	SULEV (95% cleaner than today's standard)
Performance	4-cylinder	Like a V-6

Source: Ford Motor Company. "Hybrid Vehicles," <www.ford.com/en/ourVehicles/environmentalVehicles/hybridElectricVehicles/>. Accessed October 6, 2002.

Figure 7-5   Transforming text into a table. In the original version, data is buried in the textual discussion; in the revised version, it is taken out of paragraph format and presented as a table, making it more quickly scannable and breaking up the text.

- For textual material, left-align column headings and column contents.
- For numeric material, right-align column contents and center these column contents under the column heading.
- For any narrow stream of characters (numbers, letters, symbols), left-align column contents and center these column contents under the column heading.
- Put measurement types in column or row headings, not in each of the data cells.
- To facilitate comparison of data, format tables so that data can be compared vertically. For example, cost, reliability, and technical support would be the column headings rather than the products being compared.
- Put table titles *above* tables, not below. Use the word "Table," not "Figure." Notice that table titles can be separate from the table or the first row of the table, spanning all columns.

CITE THE SOURCES OF TABLES

Whether you screen-capture someone else's table or use only portions of someone else's table, you still must document it—that is, indicate the source of that table. It's legal to copy a table verbatim from another source as long as you document its origins and as long as your document is not being sold for profit. Notice that the source of the table is indicated in Figure 7-5. However, you can also use the citation style of your documentation system—for example, the number of the source in brackets as is done in Figure 7-7.

CHARTS AND GRAPHS

The terms *charts* and *graphs* encompass the numerous ingenious ways of showing relationships between data—for example, line graphs, column charts, bar charts, pie charts, and three-dimensional variations such as pictographs. All of these types are visual representations of tables.

In tables, the significance of the data is not immediately evident. Charts and graphs, on the other hand, make that significance stand out. For example, if your department has reduced defects in the manufacturing process each year over the past five years, a line graph shows this point more vividly than a table. See also Figures 7-6 and 7-7 for illustrations of how charts and graphs can present data more dramatically than tables.

Note   See the corresponding chapter in the website companion for techniques in creating charts and graphs and dynamically incorporating them into your reports. For the web address, see the Preface.

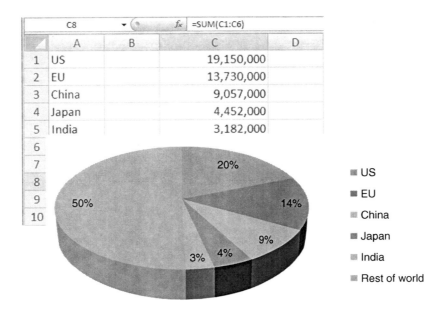

Figure 7-6   Major international consumers of oil. A spreadsheet function was used to generate the pie chart. (Thanks to Jill Brockmann for Excel assistance.)

How do you decide when to use a line graph, column or bar chart, or pie chart? Here are some ideas:

- Pie charts depict the relative portion of a total amount made up by each member that contributes to that total. Pie charts give readers a dramatic sense of the percentages of each element making up a whole. Figure 7-6 shows the top international consumers of oil.

- Line graphs depict change in data occurring over time. Several lines enable readers to compare changes between different sets of data over time. Imagine a line graph showing total sales for Dell, Hewlett-Packard, IBM, and Apple over the past decade.

Magnetic soap to clean up oil spills?

Researchers at the University of Bristol have developed a magnetic soap that can be removed from the water once it had done its job—unlike surfactants in current use which replace one form of pollution with another.

For details, see the Preface for the URL.

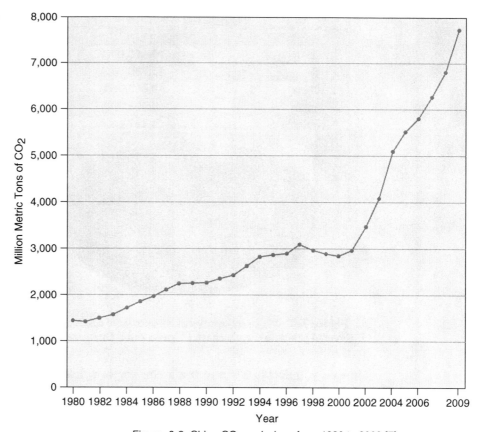

Figure 6-3. China CO_2 emissions from 1980 to 2009 [7].

Figure 7-7   Line graph depicting change over time. Notice that the title for this figure is located *below* the figure. Notice also that the source is indicated using the IEEE style of citation (see Chapter 11 for details).
Source: http://en.wikipedia.org/wiki/List_of_countries_by_carbon_dioxide_emissions

Figure 7-7 uses a line graph to show the rapid increase in China's emissions of CO_2 in recent years.

- Column and bar charts enable comparisons such as those shown in Figure 7-8, which shows carbon emissions of the top five U.S. counties. However, bar charts can reflect time as well. Imagine a chart showing bars for 1995, 2000, 2005, 2010, and 2015 for each of Dell, Apple, IBM, and Hewlett-Packard.

- Tables, on the other hand, enable the number-crunchers and the bean-counters to do their jobs. Charts and graphs are generally useless to people who have to enter precise data into electronic spreadsheets or databases.

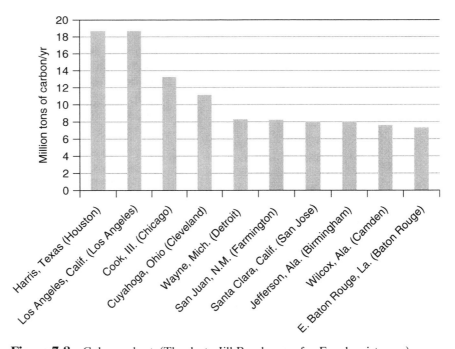

Figure 7-8 Column chart. (Thanks to Jill Brockmann for Excel assistance.)

ILLUSTRATIONS

As mentioned earlier, the term *illustration* refers to all manner of pictorial graphics—photographs, drawings, diagrams, and schematics. Included here are also conceptual diagrams such as flowcharts, even though they represent physical reality only in the most symbolic way.

TYPES OF GRAPHICS

If you must illustrate something in your engineering documents, consider carefully whether you need a photograph or some type of drawing or diagram. Photographs provide the greatest amount of visual detail which, however, may be too much. Drawings and diagrams omit unnecessary detail and enable readers to focus on the essentials. Another alternative is infographics, which have become increasingly popular. For example, Apple Inc. has published an infographic illustrating the difference between a 64-bit processor and a 32-bit processor. The difference is the size of Manhattan Island (the 64-bit processor) compared to a postcard (the 32-bit processor).

ELECTRONIC IMAGES

If you work directly with electronic images, be sure you know how to do these tasks in a graphics application such as Adobe Photoshop, Adobe Illustrator, CorelDRAW, Corel Paint Shop Pro, Open Office Draw or other similar applications:

- **Crop.** Know how to trim away unwanted material from the graphic.
- **Erase.** Know how to erase unwanted material from an image, including how to restore background to an area of erasures.
- **Combine.** Know how to combine several images into one, overlapping them if necessary.
- **Size.** Know how to enlarge or reduce a graphic, and understand the distortion that occurs when you do so.
- **Label.** Know how to add textual labels and arrows to a graphic.
- **Clean up.** Know how to sharpen, add contrast, darken or lighten a graphic.

> ### What is your ORR and ORE?
>
> In the aftermath of the Deepwater Horizon oil spill, the X PRIZE Foundation lauched an oil-spill cleanup challenge to the private sector. Taking the US$1 million first prize in 2011 was Team Elastec/American Marine with an oil recovery rate of 17,678 liters per minute and an oil recovery efficiency rate of 89.5 percent.
>
> For details, see the Preface for the URL.

GRAPHICS AND TABLES: GUIDELINES

When you incorporate graphics and tables into an engineering document, pay attention to their standard components, their placement, and cross-references to them. The following summarizes guidelines stated throughout this chapter:

- **Add figure and table titles.** Include descriptive figure titles below illustrations, diagrams, charts, and graphs. Include descriptive table titles above tables. For readers who are scanning, phrase these titles so that they identify the content of figures and tables at a glance. (You can omit titles for highly informal, obvious figures and tables.)
- **Add labels.** In illustrations, add words that identify the parts of the thing being illustrated and a pointer from each label to the part being illustrated. In charts and graphs, add labels to the axes to identify the type of information, units of measurement, and other details.
- **Indicate sources of borrowed graphics or tables.** It's easy to grab material from the World Wide Web, but remember to copy the URL, page title, any information on author and date updated, and the date *you* accessed the page.

Then include that identifying information in your document. See Chapter 11 on methods of documenting your borrowed tables and graphics.

- **Place graphics and tables at the point of first reference.** Position graphics and tables just after the first point in your text where they are relevant. If they don't fit on the same page, place them at the top of the next. Each graphic or table should appear as soon as possible after you first mention it. Place a cross-reference to the figure or table in your text *before* the figure or table occurs.

- **Align and position graphics carefully.** Maintain adequate spacing between graphics and text; make sure that graphics are nicely balanced visually on your pages. For example, if you create a graphic less than a half-page in size, you can have your text flow around it. Don't cramp things, however. Make sure you leave plenty of white space between your text and graphic and that your graphics fit within your regular margins.

- **Intersperse graphics and tables with text.** Insert graphics and tables into the main text of your document rather than appending them at the end of the document. For readers, it's pleasing to have text broken up with graphics and tables. More importantly, they need to be able to refer to the graphic or table immediately, rather than have to flip to the end of the document.

- **Include a legend.** If your graphs or charts use different symbols, colors, shadings, or patterns to indicate different elements, include a legend. See Figures 7-7 and 7-8 for examples of legends.

- **Provide cross-references to your graphics and tables.** Don't just pitch graphics and tables into engineering documents without referring to them and explaining key points. Otherwise, readers may have a nice picture or a pile of statistics, but no sense of the purpose or meaning. Use phrasing like the following:

Note As can be seen in Figure 5, the thermophysical properties...
The arrangement of the MOF network (Fig. 8-2) is structured so that...
Averages for the fabric cutting speeds are shown in Table 4 on the next page.

EXERCISES

Here are some ideas for practicing the common types of graphics and tables covered in this chapter:

1. Find a relatively simple table and reconstruct it in your own software application using the techniques and guidelines discussed in this chapter.

2. Find a relatively simple table in which the data can be converted to a line graph. Create the line graph using the techniques and guidelines discussed in this chapter.

3. Find a relatively simple table in which the data can be converted to a column or bar chart. Create the column or bar chart using the techniques and guidelines discussed in this chapter.

4. Find a relatively simple table in which the data can be converted to a pie chart. Create the pie chart using the techniques and guidelines discussed in this chapter.

5. Find text with illustrations (photographs or diagrams) on the World Wide Web, and reconstruct that page including the illustrations in your own software application using the techniques and guidelines discussed in this chapter.

BIBLIOGRAPHY

Doumont, J.L. and P. Vandenbroek. Choosing the Right Graph. *IEEE Transactions on Professional Communication, 45* (2002): 16.

Horton, William. Overcoming Chromophobia: A Guide to the Confident and Appropriate Use of Color. *IEEE Transactions on Professional Communication, 34* (1991): 160–171.

Lo, Jack and David Pressman. How to Make Patent Drawings Yourself: Prepare Formal Drawings Required by the U.S. Patent Office. Berkeley, CA: Nolo Press, 2002.

Plantenberg, K. Engineering Graphics Essentials. Mission, KS: Schroff, 2006.

Tufte, Edward R. Envisioning Information. Cheshire, CT: Graphics Press, 1990.

Tufte, Edward R. The Visual Display of Quantitative Information. Cheshire, CT: Graphics Press, 1992.

Tufte, Edward R. Visual & Statistical Thinking: Displays of Evidence for Decision Making. Cheshire, CT: Graphics Press, 1997.

8

Accessing Engineering Information

If I have seen further it is by standing on the shoulders of giants.

Sir Isaac Newton, 1642–1727.

Scientific information is growing at breakneck speed—according to some doubling every two or three years—and so are the electronic pathways to this knowledge. The information explosion is now a constant state of affairs for engineers.

Even if you work in a highly specialized field as these engineers do, you may need to access information from fields other than your own. To support you in that effort, this chapter explores engineering information resources available for your reference and research.[1]

Salmon data storage

Engineers at the National Tsing Hua University in Taiwan and the Karlsruhe Institute of Technology in Germany have created a "write-once-read-many-times" memory device that combines electrodes, silver nanoparticles, and salmon DNA, the last of which could turn out to be less expensive than traditional inorganic materials such as silicon.

For details, see the Preface for the URL.

---

[1]Many thanks to Susan Ardis, Head Librarian, Engineering Library, University of Texas at Austin, for her work on the first edition of this chapter, and to Teresa Ashley, MLS, Austin Community College, for her work on the subsequent editions of this chapter.

BASIC SEARCH STRATEGIES

Before setting out for the library or opening your favorite web search engine, know some strategies for planning and getting the most out of your search.

PREPARING FOR THE SEARCH

Although books and journals are still important sources of information (and are usually what we associate with the traditional library), they are no longer the only sources we use. The twenty-first century library is a hybrid of print and electronic resources. Since 1995, material has become increasingly and rapidly available on the World Wide Web and in libraries as well. In libraries, you can get access to subscription databases, full text of online magazine and journal articles, and growing collections of electronic books (e-books). And if you belong to such a library, your access is mostly free!

When you need information, first spend time focusing on what it is you need and where it might be. Systematically ask yourself these questions:

- What is the purpose of this information search—is it to write an internal report, work on a design problem, conduct research, or select equipment or products?
- What kind of information do you need—is it practical, theoretical, economic or public policy, proprietary, product information?
- What exactly do I need—is it raw data, an overview of the subject, historical information (for example, for product liability), up-to-date, state-of-the-art information, competitive intelligence (what's the competition up to?), intellectual property information, patents, trademarks?
- What's the time frame—hours, days, weeks, months?
- What information resources are available—nearby experts, publications that colleagues stacked away, company library, electronic access, libraries (technical, college, university, or public), technical book store?
- Are you willing to pay for the information by buying relevant books, hiring a professional searcher, paying for a full-text electronic search?

Your answers to these questions determine where you will look for information. Being as specific as possible from the beginning of your search will help you reach your information goals.

FOLLOWING THE TRAIL

When you need background or history on a subject, start down the information trail with the most readily available tools first. These are usually technical encyclopedias, handbooks, books, and periodicals.

To find specific rather than general information, be as precise as possible. Search for exactly what you want first; you can always use more general search terms later if necessary. To do that, figure out the hierarchy of your topic: what is more specific and what is more general. If your topic is photovoltaic cells, for example, that hierarchy could be any of the subdivisions shown in Figure 8-1 depending on your focus.

If you are in a library and stray off the trail, don't hesitate to ask for help. Engineering librarians often suggest that you apply the 20-minute rule: If after looking for information for 20 minutes, you find nothing relevant, ask for reference help (in most libraries, this means a trained librarian).

Another part of staying on the information trail is to become proficient with the search engines of the World Wide Web (WWW), in addition to those of library subscription databases you can access. An amazing amount of engineering data is now available from these online sources; more becomes available every day. (See "Internet Search Tools" on pages 194–196 for some starting points.) Table 8-1 provides step-by-step directions on performing electronic searches, with details on how to use keywords and Boolean operators.

Figure 8-2 is an example of searching by specific fields. Notice in Figure 8-2 that you can choose different fields within which to search, different types of documents to search, different keywords to use in the search, and different date ranges to search within. Figure 8-3 shows a login screen for Knovel Corporation which purports to be a web-based application integrating technical information with analytical and search tools to drive innovation and deliver answers to engineers.

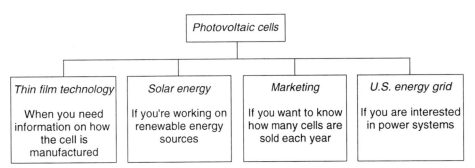

Figure 8-1 Hierarchies and subdivisions of information. In your information search, start as low in the hierarchy as you can (in this diagram, thin film technology, for example)—not high (photovoltaic cells).

Table 8-1   Searching Electronic Information Sources

To be successful in finding information online, you should feel comfortable with some of the techniques of searching electronically. Take a moment to review the techniques of searching electronically; you will more than likely be using these methods to find information.

1. Begin with a list of key terms that best describe what you hope to find. Include general, specific, and related terms.

2. Use keyword searching to locate a term *anywhere* in a record—in the author, title, publisher, subject, publisher, or other fields—or within the text of a document.

3. Use *truncation* to find the variant endings of a word. This broadens or expands your search. Insert a symbol (often an asterisk*) at the end of the root: *turbine** should retrieve both the singular and plural forms of the word.

4. Combine your keywords for more precise searching. Use the Boolean operators **and**, **or**, and **not** (sometimes **and not**) to *narrow* your search.

   and finds records that contain all your search terms. A search for *"gas turbine**"* **and** *efficiency"* would only retrieve results that contain both terms *"gas turbine**"* **and** *"efficiency"* *"Polyphenylene oxide"* **and** *"chemical resist**"*

   not finds records that contain your first search term but not your second: *energy* **not** *geothermal* would return only results that did not include the word *geothermal*.

   or broadens your search. A search for *mechanical* **or** *diesel* would retrieve results that included either *mechanical* or *diesel* or both of these terms.

5. Find out whether your search engines require that you enclose phrases in quotation marks, e.g., "gas turbine."

6. Adjacency searching lets you specify how close one term should be to another in your search results. You can use **near** for any distance within 10 characters, in any order; use **within** to specify the distance between the terms.

7. Use *parentheses* to combine different search techniques, such as combining both Boolean **and** and **or** in a single search: *CAD* and *engineer** and (*mechanical* or *electrical*). Using parentheses establishes the order in which the searches are combined and executed. Terms enclosed in parentheses are searched first.

8. Use a *field* limit to cause the system to search only the specified field for the specified word(s). Some resources provide a search page where you can check off the fields you want to include in your search.

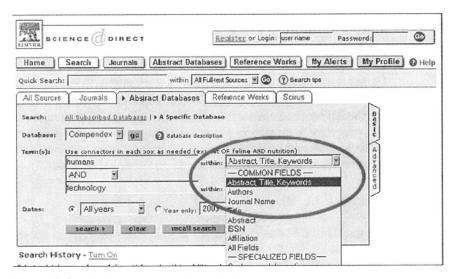

Figure 8-2 Online search page for a database.
Source: Compendex via Science Direct, www.nova.edu/library/dils/lessons/

Figure 8-3 Engineering and Scientific Online References login screen.
Source: Knovel.com search screen.

SOURCES OF ENGINEERING INFORMATION

When you search for information for an engineering project, you are likely to use an array of information resources: books, reference works, journals and e-journals (as well as the indexes and abstracts associated with them), technical reports, theses and research in progress, conference papers, patents and patent literature, standards, product information and specifications, electronic full-text sources (databases), bibliographies and reviews, Internet resources, and professional societies.

GENERAL BOOKS

In the United States alone, more than 290,000 book titles are now published annually, compared to 20,000 in 1960. Although all the information contained in these hardcopy publications could be made available electronically, this is not likely to happen for a good many years. Expect to find plenty of worthwhile information available in hardcopy only in library stacks and periodical rooms.

When to Use Books.   When you are researching a topic, books can provide excellent background, a historical treatment of your subject and depth. A quick look through a book's table of contents and index will give you a good idea of whether it's likely to have what you are looking for. eBooks provide the same kinds of information although you can use the search to quickly get to the information you seek.

For many current research topics, however, books tend to be too general. To obtain more specific information on technological advancements, you must go to journal articles, technical reports, or other sources (described later in this chapter).

Users of twenty-first-century libraries will find that library collections of eBooks are increasing exponentially. Many of the titles in your search results may be electronic and the full text will be available online. NetLibrary, a major supplier of eBooks, was purchased by EBSCOhost.com. Most college libraries provide access to this database by means of a student or patron ID. However, many other publishers make content available online as well. Here is a list of the better known sources of free online engineering eBooks:

- *The Online Books Page* at http://onlinebooks.library.upenn.edu/subjects.html. You can browse online books by subject or by Library of Congress call number.
- *Wikibooks—Subject: Engineering* at http://en.wikibooks.org/wiki/Subject:Engineering. A collection of open content textbooks that you can browse by subject, by Library of Congress call number, or by Dewey Decimal call number. These are not, however, searchable in the way that you can search the library's subscription eBook collections.

How to Find Books.   Library catalogs are online, making them accessible wherever you have access to the Internet. This means that if you cannot find a specific book or the right book in your library, you can check other libraries via the Internet. Most large public libraries and all but a few college and university libraries have online catalogs. Online library catalogs offer you powerful search tools such as those covered in Table 8-1.

See Table 8-2 for sites that list online library catalogs.

In addition to searching specific libraries, you can create a free account in OCLC's WorldCat (www.worldcat.org) and use it to search libraries worldwide. Creating a free account allows you to save your search results to a list. (Be aware that libraries subscribe to WorldCat; the free version is not as powerful as the subscription version.) You can enter a postal zip code, state, province, or country in the "Library Location"

Table 8-2   Finding Internet-Accessible, Engineering-Related Libraries

Grainger Engineering Library http://g118.grainger.uiuc.edu/voyager search/default.asp	Maintained at the University of Illinois at Urbana-Champaign, the Grainger Multiple Library Search provides an information search of Big Ten libraries and a select number of other U.S. libraries. The search can be limited to specific formats or institutions.
LibDex—www.libdex.com	Provides an index to 18,000 libraries worldwide where you can access the catalog and perform a search. The site is maintained by ITT Tech and includes links to online bookstores and other selected resources.
The WWW Library Directory www.travelinlibrarian.info/libdir	Provides a list of lists of libraries worldwide that may duplicate others in this list.
LibWeb www.indiana.edu/˜librcsd/internet/ libweb-mirror	Maintained at the University of California, Berkeley, this resource lists over 7,700 pages from libraries and enables you to search libraries in over 146 countries.
National Libraries and Research Centers: The Library of Congress Catalog http://catalog.loc.gov The British Library www.bl.uk Canada Institute for Scientific and Technical Information (CISTI) http://cisti-icist.nrc-cnrc.gc.ca/cisti_e.html	These are three of the world's major sources of information in all areas of science, technology, engineering, and medicine. Materials are collected worldwide and in various languages. You may find links to other libraries' catalogs as well as catalogs to these institutions' resources.

search box to find the closest local library owning the title you need. Once you find a book that your library does not own, use the FindIt! button to locate the library and request an Interlibrary Loan. The Shapiro Library provides a video tutorial on using WorldCat to find books: www.youtube.com/watch?v=2VrxPjDjw-U.

Google Books (http://books.google.com), Yahoo!, and other web search engines can access the public WorldCat database. For more precise searching, include either of the following key phrases with your search terms: "find in a library" (include quotation marks) or add site:worldcatlibraries.org (no space after colon) after your search terms.

Search for a specific topic on the Google Books website. You can use Advanced Book Search to apply limits to your search and make it more precise. If, for instance, you want to search for particular words in the title, either use the "Title" search box in Advanced Book Search or precede your search with "intitle:" As an example, this search: **intitle: engineering ecology** would retrieve *Ecological Engineering: Bridging Between Ecology and Civil Engineering.*

> **Battling terabyte SSDs: Part I**
>
> In the fall of 2011, OCZ released the world's first terabyte 2.5-inch solid state disk. (Terabyte, that's 10^{12}.)
>
> For details, see the Preface for the URL.

How to Obtain Books. What if you find something in a library 400 miles away? If you have time and don't plan to travel to that location, ask your librarian about interlibrary loan. (If waiting for the book to arrive by mail is not an option, at least you know what is available on your topic.)

REFERENCE WORKS

In addition to books located in the library stacks or available online as eBooks, most engineering libraries (and technical information centers) have a reference section where you can find books that provide quick answers to questions or specific facts, such as the molecular weight of a compound. Although we still speak of these types of materials as "reference books," we might more properly refer to them as "reference works" or "reference sources" since an increasing number of these resources are electronic equivalents like CD-ROMs, eBooks, and databases where the content is updated frequently, sometimes daily. Generally, the types of works that fall into the "reference" category are abstracts and abstracting services, almanacs, bibliographies, dictionaries, directories, encyclopedias, guidebooks, handbooks, indexes, manuals, yearbooks, and similar materials that are usually consulted rather than read cover to cover. If they are bound, these works normally cannot be taken out of the library building but are used for quick, on-the-spot "look-up" of factual information. If they are online, they are usually available through a library subscription and accessed by user login and password.

The best ways to find useful reference tools are to ask a librarian for assistance or to use the online library catalog. Using the keyword search option, type in broad terms like "*engineer** and *dictionary*" or "*engineer** and *encyclopedia*" (the * asterisk truncates the term so you can find variant endings, for example, "*engineer, engineers, engineering*). If you don't know the URL (web address) of a library's catalog, just type in the name of the institution in quotes and "library catalog" in the search field of Google (www.google.com), Yahoo! (http://search.yahoo.com/) or your favorite general Internet search engine. (See the "Internet Search Tools" section for a list of tools.)

When to Use Reference Books. As you can see from Table 8-3, reference works include a broad range of engineering reference books. The table lists just a sampling of the wide range of reference books and their online equivalents that engineers might consult.

Because engineers must be knowledgeable about areas beyond their engineering specializations in a globalized work environment, reference works can be a good starting point or foundation for what is known in a field and are especially useful for cross-disciplinary information searches.

Any of the reference titles listed earlier can be found by typing in a couple of words of the title in an online library catalog (for example, *mechanical engineer** handbook*, "*encyclopedia engineering*," or "*encyclopedia* **and** *engineering*"), truncating any words that could have variant endings, and eliminating any prepositions or articles (*of, for, the, a, an*).

How to Find Reference Books. Like all books in the library, reference books are shelved according to the classification system that is in use. Most college and university libraries use the Library of Congress classification system (LC) and may also use SuDocs (Superintendent of Documents Classification) numbers for government documents; public libraries and some technical libraries may use the Dewey Decimal Classification system (DDC). If the reference work is electronic, you may need to get help to find the path to the online resource. Some online reference works are in aggregated collections of reference titles and several clicks of the mouse can get you to the title and pages you need. Electronic resources usually provide retrieval options that include printing the information, emailing it to yourself, or saving it to a computer storage device like a flash drive.

JOURNALS

You probably already subscribe to one or two professional journals and may have access to others through a local library. Over 10,000 hardcopy and hundreds of electronic scientific and technical journals are published every year, and both numbers are growing. Journals are essential for any engineer who wants to keep up with latest developments. Journals, as opposed to popular magazines, contain scholarly, "peer-reviewed" articles. *Peer-reviewed* means that articles have been evaluated by professionals (peers) and experts in the field, prior to publication in the journal. Peer-reviewed articles are sometimes called "refereed."

Table 8-3 Selected Examples of Reference Works

CRC Handbook of Mechanical Engineering. 2004. Contains useful articles, tables, and data on all aspects of mechanical engineering and other subjects of use to mechanical engineers. It is also available as an eBook. Similar works include: *Handbook of Industrial Engineering*, 2007, and *The CRC Handbook of Thermal Engineering*, 2000.

Encyclopedia of Energy Technology and the Environment, 1995. Part of the Wiley *Encyclopedia Series in Environmental Science.* Four volumes of articles on energy-related topics relating to technology and its impact on the environment.

ENGnetBASE (www.crcnetbase.com/page/engineering_ebooks). An online database with the full-text of over 75 engineering handbooks.

The Kirk-Othmer Encyclopedia of Chemical Technology, 5th ed., 2006. 26 volumes. Covers all areas of technology—not just chemical. At the end of each article are useful references to patents, conference proceedings, and journal articles.

The Kirk-Othmer Encyclopedia of Chemical Technology Online (http://onlinelibrary .wiley.com/book/10.1002/0471238961). Updated regularly. Searchable, full text access.

McGraw-Hill Dictionary of Scientific and Technical Terms, 6th ed., 2002. Provides more than 125,000 definitions of terms and includes some 3,000 illustrations.

The McGraw-Hill Encyclopedia of Science and Technology, 11th ed., 2012. 20 volumes. Contains almost 8,000 well-written and well-illustrated articles on science, engineering, and other technical subjects. Check here first for general background information.

AccessScience @ McGraw-Hill: The *Online Encyclopedia of Science & Technology.* The online version of *The McGraw-Hill Encyclopedia of Science and Technology*; includes dictionary terms, Research Updates, and other resources. Updated daily.

Knovel: Engineering & Scientific Online References An aggregate online collection that provides electronic versions of standard reference books in engineering and applied sciences. Includes *Perry's Chemical Engineers' Handbook* and many other works.

Marks' Standard Handbook for Mechanical Engineers, 11th ed., 2006. Continues the *Standard Handbook for Mechanical Engineers.* The 10th edition is available online by subscription at www.knovel.com and may be in some engineering library collections.

Perry's Chemical Engineers' Handbook, 8th ed., 2007. Includes material from general mathematics and tables to specialized treatment of topics such as psychometry, process machinery, and distillation. A standard for petroleum and chemical engineers.

Standard Handbook for Civil Engineers, 5th ed., 2003. Covers construction, structural theory and design, materials, and management for the various fields of civil engineering, including environmental concerns.

Standard Handbook for Electrical Engineers, 15th ed., 2006. Substantial coverage of all aspects of electrical engineering, with numerous tables, charts, and graphs.

Van Nostrand's Scientific Encyclopedia, 10th ed., 2008. 2 volumes. Concentrates on the basic and applied sciences, with over 17,000 articles. Also functions as a technical dictionary.

When to Use Journal Articles. The information in journals (unlike books) is highly technical and contains current research in an area of specialization. Few libraries can subscribe to more than a fraction of the journals published, although library subscriptions to journal databases with full-text articles have greatly increased the depth and breadth of library holdings. The limitations of a particular library's collection can be overcome by interlibrary loans, of course.

How to Find Journal Articles. To become familiar with all the journals published in your field, consult *Ulrich's International Periodicals Directory*, which annually lists journal titles in some 200 categories, including engineering, which itself is further subdivided by fields such as civil, electrical, mechanical, and petroleum engineering. Many public, college, and university libraries own *Ulrich's* or have access to it by subscription to the online version at www.ulrichsweb.com.

You may also want to take a look at the Directory of Open Access Journals (www.doaj.org). DOAJ offers free access to over 3,500 full-text, quality-controlled scientific and scholarly journals, over 1,200 of which are searchable at the article level. As part of the open access movement, DOAJ aims "to increase the visibility and ease of use of open access scientific and scholarly journals thereby promoting their increased usage and impact." You can browse by title or by subject. Engineering titles can be found under "Technology and Engineering" with further subdivisions by field of engineering.

INDEXES AND ABSTRACTS

If you intend to use research articles for an engineering document, be aware of two essential tools for finding and selecting those articles: indexes and abstracts. Increasingly, these are one and the same, as online indexes usually include abstracts or summaries of articles.

Indexes. Imagine that you wanted to find all the research articles written on fuel cells before President George W. Bush announced his intentions for the Department of Energy to refocus its work on that technology in 2002. How would you find all those articles? Spend weeks scanning the table of contents of likely journals? No, instead you'd use a *periodical index* to find articles on fuel cells published in a wide array of magazines and journals.

An index lists articles grouped by subjects from selected periodicals. For each article, you'll find the article title, author (or authors), periodical title, volume, issue, date, page numbers, and page count. Indexes are available for broad categories or fields, as shown in Table 8-4.

Abstracts. Periodical indexes enable you to find articles on a specific topic. If you find too many articles, read the *abstract* of each article to decide whether to use that article. Abstracts appear with the articles themselves. But you can also look at just the abstracts without having to go to the articles themselves. Most libraries provide

Table 8-4   General and Engineering-Specific Indexes

General indexes that cover engineering as well as other disciplines:

- **Academic OneFile.** Peer-reviewed, full-text articles with extensive coverage of the physical sciences.
- **Academic Search Complete.** Provides full text for nearly 5,990 journals, including full text for more than 5,030 peer-reviewed titles.
- **Applied Science and Technology.** Indexes articles, product evaluations, and book reviews in over 390 English-language periodicals; subjects covered include all fields of engineering.
- **IngentaConnect.** A comprehensive multidisciplinary document delivery service; articles are available in downloadable electronic format or as fax documents.
- **ISI Web of Science.** An online version of the Institute for Scientific Information's citation indexes, including the *Science Citation Index*.
- **INSPEC.** A database offering physicists, engineers, computer scientists, and information specialists access to international journal articles, conference proceedings, reports, dissertations, and books covering physics, electronics, information technology, computers, and electrical engineering.
- **ScienceDirect.** Provides access to more than 1,000 periodicals (over 1,100,000 articles).

Indexes specifically for engineering:

- **Compendex.** Online version of the Engineering Index, and the most comprehensive interdisciplinary engineering database in the world, referencing 5,000 engineering journals and conference materials dating from 1970.
- **IEEE Xplore.** Covers more than 30 percent of the world's literature in electrical engineering, electronics, materials science, physical sciences, and biomedical engineering, with full text access to over 140 journals.

a list of their electronic databases and online periodical indexes on the library home page. For an example, see the listing provided by Stanford at http://library.stanford .edu/sulapp/databases/index.jsp?function=search&field=discipline&query=Science%20 and%20Engineering. Look particularly at the indexes and abstracts marked "Open to all." For example, select **Astrophysics Data System Abstract Service** and then **Astronomy and Astrophysics Search.** Figure 8-4 shows part of the search interface; Figure 8-5 shows what one of the index entries looks like, along with its abstract.

Unfortunately, few electronic indexing and abstracting services are "open to all" as is the Astrophysics Data System Abstract Service. You cannot access these resources unless you are affiliated with an institution that pays for (subscribes to) them, such as an academic or corporate library.

Enter Abstract Words/Keywords   ☐ Require text for selection
(Combine with: ⦿ OR ◯ AND ◯ simple logic ◯ boolean logic)

potatoes spinach wheat tadpoles fish mars space

Return ☐100 items starting with number ☐1

[Send Query]   [Return Query Form]   [Store Default Form]   [Clear]

Figure 8-4 Search terms entered into a typical search interface. This would be a search for research articles on the effects of extraterrestrial existence on animal and plant life.
Source: Search terms entered at Astrophysics Data System Abstract Service, Astronomy and Astrophysics Subcategory, http://library.stanford.edu/sulapp/databases/index.jsp?function =search&field=discipline&query=science%20and%20engineering

Title:	Effects of modified atmosphere on crop productivity and mineral content
Authors:	Chagvardieff, P.; Dimon, B.; Souleimanov, A.; Massimino, D.; Le Bras, S.; Péan, M.; Louche-Teissandier, D.
Affiliation:	CEA, Direction des Sciences du Vivant, Départment d'Ecophysiologie Végétale et de Microbiologie, Centre de Cadarache, F-13108 Saint-Paul-Lez-Durance cédex, FRANCE
Journal:	Advances in Space Research, Volume 20, Issue 10, p. 1971-1974. (AdSpR Homepage)
Publication Date:	00/1997
Origin:	ELSEVIER
Abstract Copyright:	(c) 1997 Elsevier Science B.V. All rights reserved.
Bibliographic Code:	1997AdSpR..20.1971C

Abstract

Wheat, potato, pea and tomato crops were cultivated from seeding to harvest in a controlled and confined growth chamber at elevated CO_2 concentration (3700 muL.L^-1) to examine the effects on biomass production and edible part yields. Different responses to high CO_2 were recorded, ranging from a decline in productivity for wheat, to slight stimulation for potatoes, moderate increase for tomatoes, and very large enhancement for pea. Mineral content in wheat and pea seeds was not greatly modified by the elevated CO_2. Short-term experiments (17 d) were conducted on potato at high (3700 muL.L^-1) and very high (20,000 muL.L^-1) CO_2 concentration and/or low O_2 partial pressure (~ 20,600 muL.L^-1 or 2 kPa). Low O_2 was more effective than high CO_2 in total biomass accumulation, but development was affected: Low O_2 inhibited tuberization, while high CO_2 significantly increased production of tubers.

Figure 8-5 Abstract—example. This abstract is typical of what you see in electronic indexing and abstracting services. You get both the index entry with bibliographic detail to enable you to find the complete article, plus the abstract, which provides a summary of the research purpose and outcomes.

How to Obtain Journal Articles. Not all scientific and engineering articles can be retrieved through traditional bibliographic sources such as periodical indexes. You can also search online collections of preprints (also called "e-prints"), particularly popular in the sciences. Although preprint articles have not yet been published, they may have been submitted, reviewed, and accepted for publication or made available online prior to presentation at conferences. Preprints are of interest because research findings are available sooner than through the traditional peer-review process. Here are two resources for finding preprint articles:

- The ArXiv e-print archive: Articles in physics, mathematics, nonlinear science, and computer science, operated by Los Alamos National Laboratory, accessible at http://arxiv.org
- E-preprint Network: Research Communication for Scientists and Engineers operated by the Department of Energy (DOE) Office of Scientific and Technical Information (OSTI), accessible at www.osti.gov/eprints

TECHNICAL REPORTS

Hundreds of thousands of technical reports are written each year in the United States alone; many are available on electronic media. A technical report may be similar to a paper presented at a conference or to a journal article, but it may be a lot longer. Technical reports are usually written by specialists for other specialists; they typically report on the results of research and development.

Reports sponsored by a government grant or contract are the easiest to find, whereas proprietary and classified reports are not generally available. To find them, you must use indexes and abstracts such as NTIS and NASA RECON, described in Table 8-5. (See Figure 8-6 for an illustration of a typical NTIS record.)

PATENTS

Patent documents describe in detail the designs, materials, machines, and processes associated with inventions. In return, the government grants the inventor a right of ownership that limits others from making, using, or selling the patented item in the United States for 20 years. Currently, the number of U.S. patents granted annually is close to 180,000. Since the 1790s, some 7,950,000 had been granted as of February 2008. As Table 8-6 shows, it's amazing what gets patented.

Battling terabyte SSDs: Part II

But just a few months later after OCZ's 1 terabyte SSD, Victorinox released an SSD of the same storage capacity but fitted nicely into a Swiss army knife.

For details, see the Preface for the URL.

Table 8-5 Finding Engineering Reports

National Technical Information Service (NTIS) www.ntis.gov	Major source for information on nonproprietary and unclassified reports sponsored by government agencies and contractors. You can search technical reports on government-sponsored research from organizations such as NASA, DOE, and EPA; read abstracts for the reports; and make online purchases.
NTRS-NASA Technical Reports Server http://ntrs.nasa.gov/search.jsp	NASA's STI (scientific and technical information) research reports, journal articles, conference and meeting papers, technical videos, mission-related operational documents, and preliminary data. Available via the NASA Technical Report Server (NTRS).
NASA Technical Reports Server http://ntrs.nasa.gov/search.jps	Access to approximately 500K aerospace related citations, 90K full-text online documents, and 111K images and videos as well as access to NASA's aerospace research and engineering results.
IEEE Xplore http://ieeexplore.ieee.org/ Xplore/guesthome.jsp	Access to the nearly 2 million documents available from IEEE, including reports, journals, transactions, and magazines, IEEE conference proceedings, IEEE journals, IEEE conference proceedings and Current IEEE standards, all published since 1988. (Available through subscription.)

If you've never seen a U.S. patent, look at Figure 8-7. Each front page includes the inventor's name (patentee), owner at date of issuance (assignee), date issued, citations to other relevant patents and articles, one drawing, and an abstract. Following the front page is a disclosure section, where inventors describe or "disclose" how their inventions work and how they relate to or improve on existing solutions to the same problem, and a claims section where inventors give the legal description of what is actually protected by the patent.

When to Use Patent Information.   In your professional work, do a patent search when you want to:

- Find out about a specific patent.
- Learn about recent inventions in a particular field.
- Find out if your invention has already been patented.
- Gain ideas for further development of your invention.
- See what inventions known competitors have patented.

You may be unaware of the enormous amount of technical information contained in patent documents. In fact, you cannot find descriptions of most of the technology

```
1824205 NTIS Accession Number: N95-16175/2/XAB
    Simulation of the Coupled Multi-Spacecraft Control Testbed at the
Marshall Space Flight Center
    Ghoah, D. ; Montgomery, R. C.
    National Aeronautics and Space Administration, Hampton, V&. Langley
    Research Center.
    Corp. Source Codes: 019041001; ND210491
    Oct 94 22p
    Languages: English
    The Role of Computers in Research and Development at Langley
    Research Center p. 497-517.
    NTIS Prices: (Order as N95-16453/9, PC A99/MT A06
    Country of Publication: United States
    The capture and berthing of a controlled spacecraft using a robotic
manipulator is an important technology for future space missions and is
presently being considered as a backup option for direct docking of the
Space Shuttle to the Space Station during assembly missions. The
dynamics and control of spacecraft configurations that are manipulator-
coupled with each spacecraft having independent attitude control systems
is not well understood and NASA is actively involved in both analytic
research on this three dimensional control problem for manipulator
coupled active spacecraft and experimental research using a two
dimensional ground based facility at the Marshall Space Flight Center
(MSFC). This paper first describes the MSFC testbed and then describes a
two link arm simulator that has been developed to facilitate control
theory development and test planning. The notion of the arms and the
payload is controlled by motors located at the shoulder, elbow, and wrist.

    Descriptors: • Attitude control; • Computerized simulation; • Control
theory; • Dynamic control; • Manipulators; • Robot arms; • Space shuttles;
• Spacestations; • Spacecraft configurations; • Spacecraft control;
• Spacecraft docking; Equations of motion; Ground tests; Payloads;
Robotics; Shoulders; Space missions; Wrist
    Identifiers: HTISMASA
    Section Headings: 84A (Space Technology--Astronautic«)
```

Figure 8-6 Typical record available from NTIS (National Technical Information Service). Notice that the paper described is part of an internal Langley Research Center report and that the entire report must be purchased. Reports labeled with a PC (or price code) can be ordered from NTIS (1-800-336-4700) on paper or microfiche.

in U.S. patent information in any other source. Because patent searching is complex, read about the process in one of these general sources:

- Timothy Lee Wherry. *The Librarian's Guide to Intellectual Property in the Digital Age: Copyrights, Patents, and Trademarks.* Chicago: American Library Association, 2002.

- David Hunt, Long Nguyen, and Matthew Rodgers, eds. *Patent Searching: Tools & Techniques.* New York: John Wiley, 2007.

Table 8-6 Famous and not-so-famous patents—examples. Put the patent number in the search field at http://patft.uspto.gov/netahtml/PTO/srchnum.htm and take a look

Coca-Cola Company. *Design of the bottle.* Patent number 696,147

William M. Mirick. *Correction fluid composition.* Patent number 3,674,729

Mervin R. Williams. *Illuminated hula hoop.* Patent number 4,006,556

Lynda S. Samen. *Combined earthquake sensor and night light.* Patent number 4,978,948

Yau, Chiou C., et al. *Ozone-friendly correction fluid.* Patent number 5,199,976

Aaron Harrell. *Pneumatic shoe lacing apparatus.* Patent number 5,205,055

W. Roelofs, et al. *Cockroach attractant.* Patent number 5,296,220

Israel Siegel. *Gravity powered shoe air conditioner.* Patent number 5,375,430

David Falco. *Versatile necktie tying aid gauge.* Patent number 5,505,002

F. Robert Egger. *Bicycle helmet.* Patent number 5,651,145

Dean L. Kamen, et al. *Human mobility vehicle.* Patent number 6,367,817

- David Hitchcock. *Patent Searching Made Easy.* 5th ed. http://Lulu.com, 2007.
- *General Information Concerning Patents: A Brief Introduction to Patent Matters.* U.S. Department of Commerce, Patent and Trademark Office. Washington, DC: Author, 1992.
- Timothy Wheery. *Patent Searching for Librarians and Inventors.* Chicago: American Library Association, 1995. Explains important differences between copyrights, patents, and trademarks.

You can learn more about patents and patent searching by visiting these sites:

- University of Texas, McKinney Engineering Library. Patent Searching Tutorial: www.lib.utexas.edu/engin/patent-tutorial/tutorial/pattut.html
- Penn State University, Schreyer Business Library's Patent Search Tutorial: www.libraries.psu.edu/instruction/business/Patents/index.html
- Brian Mathews. *Patent Searching for ME: Part 1.* Georgia Tech Mechanical Engineering. YouTube, July 22, 2006. www.youtube.com/watch?v=pQmu E5kzkzo

How to Find Patent Information. The best place to find recent patent information is the *Official Gazette* of the United States Patent and Trademark Office (USPTO) at www.uspto.gov/go/og/index.html. The *Official Gazette* contains brief descriptions and drawings of the some 1,500 patents granted every Tuesday. To get to the USPTO patent-search page, go to http://patft.uspto.gov. A more efficient way to search patents is to go to a Patent and Trademark Depository library (PTDL), a list of which can be found at www.uspto.gov/go/ptdl. PTDLs provide free access to other search tools.

US006043842A

United States Patent [19]
Tomasch et al.

[11] **Patent Number:** 6,043,842

[45] **Date of Patent:** Mar. 28, 2000

[54] **REMOTE SURVEILLANCE DEVICE**

[75] Inventors: **Michael D. Tomasch**, Massapequa Park, N.Y.; **Anthony G. Martin**, Trabuco Canyon, Calif.

[73] Assignee: **Olympus America, Inc.**, Melville, N.Y.

[21] Appl. No.: **08/775,311**

[22] Filed: **Dec. 31, 1996**

[51] **Int. Cl.**[7] H04N 7/18
[52] **U.S. Cl.** **348/164**; 348/143; 385/118
[58] **Field of Search** 348/164, 143, 348/65, 68; 128/898; 385/118; 73/864.73; H04N 7/18

[56] **References Cited**

U.S. PATENT DOCUMENTS

Re: 33,572 4/1991 Meyers

4,027,159	5/1977	Bishop
4,261,204	4/1981	Donaldson 73/864.73
4,574,197	3/1986	Kliever
4,696,544	9/1987	Costella 385/118
4,707,595	11/1987	Meyers
4,998,282	3/1991	Shishido 381/77
5,130,527	7/1992	Gramer et al.
5,215,105	6/1993	Kizelshteyn 128/898

Primary Examiner—Howard W. Britton
Attorney, Agent or Firm—Michaelson & Wallace; Peter L. Michaelson; John C. Pokotylo

[57]   **ABSTRACT**

A remote surveillance system including an imaging device and an IR light source for surveying a relatively dark area, a remote surveillance system including an imaging device and an insertion tube guide, and an insertion tube guide for receiving an insertion tube of an imaging device.

47 Claims, 12 Drawing Sheets

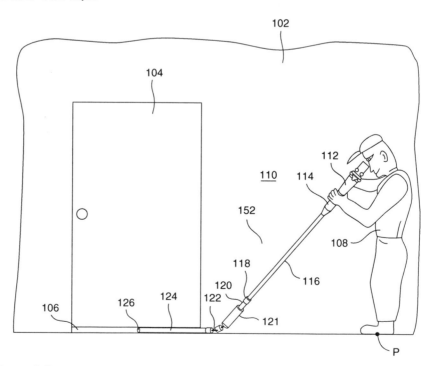

Figure 8-7 Front page of a U.S. patent document. Averaging ten pages, patent documents also include disclosure and claims sections as well as interesting drawings like this one.

For information on patents issued in other countries, the best sources are online. The cost varies widely. Three useful sources are:

Japan Patent Information Organization (JAPIO)	www.japio.or.jp/english
International Patent Documentation Center (INPADOC)	www.epo.org/searching/subscription/raw/product-14-11.html
Derwent World Patents Index (DWPI^{SM}) (DERWENT)	database of international patent documents

If you are interested in applying for a patent for your own work, begin with these two good resources:

- David Pressman, *Patent It Yourself.* Berkeley, CA: Nolo Press, 15th ed. 2011. (Text and software are available.)
- Susan Ardis, *Introduction to U.S. Patent Searching.* Westport, CT: Libraries Unlimited, 1991.

PRODUCT LITERATURE

A gold mine of information for engineers can be found in product literature, which includes product, manufacturer, company, and vendor catalogs, product selectors, buyer guides, and so on. You'll find performance data, photographs or drawings of products, data books for computers and integrated circuit devices, application notes, and other information about specific products. Topics can range from aerospace ordnance equipment to transportation and vehicle equipment or supplies. Sales representatives and most libraries can help you, or you can use a search engine to find the company website.

When to Use Product Literature. If you are on a design project, product literature is indispensable. You can get the dimensions or performance figures for specific components, accessories, or equipment related to your project. Using these resources enables you to know what is already available in your field and to compare currently available products.

An individual product catalog usually features just one product, while manufacturer catalogs show a variety of products for sale from a specific manufacturer (see Figure 8-8). Vendor catalogs, on the other hand, show all products for sale by the vendor and are designed for fast and easy comparison between several competing brands. Any of these usually provide at least limited product specifications, performance data, drawings, test data, and application details.

One example of a product catalog dedicated to a specific field is the *Electronic Engineer's Master* (EEM) catalog. Designed to help users see a range of similar products, EEM is a collection of pages from catalogs of companies around the world. The online edition can be found at www.eem.com

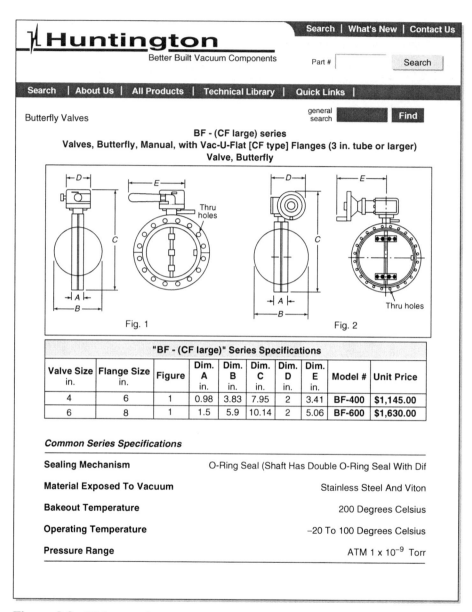

Figure 8-8 Web page from Huntington Laboratories' online catalog, Huntington Mechanical Laboratories, Inc., 2003. www.huntvac.com (reprinted by permission).

How to Find Product Literature.   To get an idea of the enormous variety of products described in catalogs, look at the annual *Thomas Register of American Manufacturers*, commonly called *ThomCat*. The first sixteen volumes provide access to the names, addresses, and telephone numbers of about 150,000 U.S. manufacturing firms. You can look up companies that make a specific product—for example, backhoes. Other volumes of interest include a U.S. tradename index; you can use this index to find out who manufactures Teflon, for example. Volumes 19–26 consist of selected pages from individual company catalogs.

The online version of *ThomCat* is at www.thomasnet.com/index.html. A search field, with a dropdown menu of state options will help you narrow your search. You can select tabs to get a list of links to various product categories, company catalogs, or CAD drawings.

STANDARDS AND SPECIFICATIONS

Most products we use daily are designed and produced in accordance with certain standards or specifications. The length of toothpicks, the softness of toilet paper, and the different grades of sandpaper are all controlled by agreed-upon industrial standards. These standards are essential if you want to be able to consistently fit light bulbs into sockets, screw nuts onto bolts, replace engine parts, or rely on the strength of concrete.

When to Consult Standards and Specifications.   As a design engineer, you must be aware of what standards, specifications, or codes already exist that might be relevant to your product. One professional engineer puts it this way:

> *You will usually be informed of the applicable specs by your managers, but they may miss some. If you do not comply with an applicable spec, you will have to redesign, with cost to your organization and criticism of yourself whether or not it was your fault. It is good practice to assure yourself independently that you know all applicable specs.*

> Lawrence L. Kamm, *Successful Engineering: A Guide to Achieving Your Career Goals* (New York: McGraw-Hill, 1989), p. 145.

The standards for specific products are set by the trade associations, companies, manufacturers, and professional societies involved in those products, and also by government agencies and international standards organizations. Table 8-7 shows a few of the hundreds of organizations that produce standards.

How to Find Standards and Specifications.   Here are some efficient ways to locate standards information:

- American National Standards Institute (ANSI). *Catalog of American National Standards* (traditional printed resource). Website: www.ansi.org
- Department of Defense (DOD). *Index of Specifications and Standards* (traditional printed resource). Federal product descriptions can be downloaded on the Internet with no registration required at www.dsp.dla.mil. Click on "Online

Table 8-7   Engineering-Related Electronic Discussion Lists—Examples

CAEDS-L	Computer Aided Engineering Design (CAEDS) Interest Group.
CHEME-L	Covers the role of chemical engineering in technology and world economies and serves as an open forum for various technical, professional, and educational issues.
CIVIL-L	Civil Engineering Research & Education.
ENVENG-L	For those interested in education, research, and professional practice relating to environmental engineering.
MATERIALS-L	For those involved in both teaching and research in materials science and engineering.
MECH-L	For discussions of mechanical engineering, including composite materials and other ME-related topics.
METALLURGY-L	Covers all aspects of metallurgical engineering, including (but not necessarily limited to) mineral processing, extractive metallurgy, hydrometallurgy, pyrometallurgy, metals refining, alloying, welding, casting, and metallography.
TDR-L	Discussion of Time Domain Reflectometry issues for engineering and geo measurements.
Intute: science, engineering, and technology	Search for *engineers* "*discussion lists*" or find E-mail Discussion Lists at www.intute.ac.uk/sciences/cgi-bin/browse.pl?id=489

Specs"; then click "Click here to get DSP documents"; Click on "ASSIST-Quick Search."

- **Information Handling Service (IHS).** *Industry Standards and Engineering Data: Number and Subject Index* (general tool covering multiple organizations).
- **National Standards Association.** *Standards and Specifications* (government and industry standards and specifications including FEDSpecs and MILSpecs).
- **International Telecommunications Union (ITU).** Recommendations: www.itu.int/net/home/index.aspx
- **International Organization for Standardization.** ISO Online: www.iso.org/iso/home.htm

Figure 8-9 shows the typical structure of a standard. In this brief example, pay attention to the parenthetical "R 1993." This indicates that the issuing organization reviewed and reapproved this 1983 standard in 1993. Always check to be sure you are using the most up-to-date standard.

U.S. Government Specifications

As one of the world's largest buyers of practically every kind of civilian and military product or service, the U.S. government has produced enormous quantities of standards and specifications describing requirements for its purchases. In 1995, the U.S. Congress

```
Resolution:
    Used for:    Limiting Spatial Resolution: LSR; Resolving Power
    See also:    Image Intensifiers: Numerical Aperture
Tolerances
    ANSI PH3>609-80 Dimensions for Resolution Test Target for
    Photographic Optics (R 1987) NFP(A) T2.9.6 R1-90. Hydraulic Fluid
    Power--Calibration Method for Liquid Automatic Particle Counters.
    (Revision/Re designation ANSI B93.28-1973)
Cathode-Ray Tubes
    EIA TEB25-85 Survey of Data-Display CRT Resolution Measurement
    Techniques.
Copying Machines
    ASTM F807-83 Standard practice for Determining Resolution
    Capability of Office Copiers. (R 1993)
Image Processing Systems
    AIIM TR26-93 Resolution as it Relates to Photographic and
    Electronic imaging.
Photographic Lenses
    ANSI PH3.63-74. Method for Determining the Photographic Resolving
    Power of Photographic Lenses. (R 1991)
```

Figure 8-9 Description of an industrial standard from *Industry Standards and Engineering Data: Subject Index.*

and the GAO instructed government agencies to use ASME, ASTM, ANSI, IEEE, UL, and other standards or specifications whenever possible. Specifications for items bought by the General Services Administration and the U.S. armed forces are as follows:

- United States Federal Standards and Specifications (FEDSPECS)
- United States Military Standards and Specifications (MILSpecs)
- Department of Defense Index of Specifications and Standards (DODISS)

Battling terabyte SSDs: Part III

An engineer at the Institute of Materials Research and Engineering in Singapore has found that adding sodium chloride (table salt) to a developer solution used in existing lithography processes can increase the capacity of a hard disk drive to 6 terabytes.

For details, see the Preface for the URL.

Some engineering libraries have IEEE, ANSI, and ASME standards collections. Those with large collections may also provide access to one of the electronic indexes mentioned above. Commercial vendors, such as Global (1–800–624–3974), provide express-service copies of U.S. and international standards.

INTERNET ENGINEERING INFORMATION RESOURCES

To this point, you have read mostly about print-based methods of finding information, some of which have electronic analogs, such as *Engineering Index* online. However, the Internet itself offers some other alternatives for information research to the engineer.

ELECTRONIC MAILING LISTS OR DISCUSSION GROUPS

Mailing lists are ongoing Internet-based discussions focused on a particular topic and used by people all over the world. Mailing list activity is simply a series of email messages, written rapidly by anyone with or without any qualifications and all strung together. Despite their questionable reliability and fragmented nature, mailing lists are a good way to get information and opinions and, more importantly, names and addresses to contact.

Here's how a mailing list works: Let's say you are a civil engineer and want to enter into discussions with other civil engineers worldwide. You "subscribe" to CIVIL-L; thereafter, you receive any email that anyone else subscribed to CIVIL-L sends to that list; any email you send to the list gets sent to all the other subscribers to CIVIL-L. Many, but not all, of these electronic mailing lists "archive" their email activity. Not only can you watch current email for your topic, you can search these archives for your topic and see what subscribers have said about it. Table 8-7 provides a sampling of the many mailing lists that are available.

Although it takes some effort to subscribe to a list, postings on a mailing list or discussion list are likely to be more thought through by their authors and more respected by their readers. Still, messages tend to be written hastily and merely offer opinions, and you cannot be sure of the senders' qualifications. Even so, the information and contacts you can get from following a mailing list or rummaging in its archives can be invaluable, even if you don't trust the information itself.

How do you find discussion lists in your areas of interest? Try the possibilities shown in Table 8-8.

Table 8-8   Finding Engineering-Related Electronic Mailing Lists

CataList, the Official catalog of LISTSERV® lists www.lsoft.com/lists/listref.html	Browse or search the 72,443 public LISTSERV lists on the Internet, read brief descriptions, subscribe.
Tile.Net/Lists www.tile.net	Browse by name, description, or domain.
Google—www.google.com Yahoo!—http://search.yahoo.com	Use a search engine to search for *engineers "discussion lists"* in the search box.

BLOGS

Blogs are like online diaries or journals. They don't have the same interaction as a discussion list, although readers can leave posts in response to discussions chosen and led by the blog owner. Consider that engineers often work for companies where there is proprietary information, which limits what can be shared with others. While engineers may use online communication for problem solving, there may be more opinion and news traded than technical information. That said, Table 8-9 lists a few engineering blogs that might serve as current awareness sites.

To find blogs, use www.icerocket.com or another web search engine. If you type "mechanical engineering" in the search box, you will probably retrieve quite a few posts with those terms. Unfortunately, many of them may be job postings.

ELECTRONIC NEWSLETTERS

Newsletters are less formal publications than journals, which report research findings. Professional societies may publish newsletters, some of which are freely available on the Web. Access to these online-only periodicals varies widely. Some you subscribe to; some are free; some you pay for; they arrive weekly, monthly, or quarterly in your email while others you access online from sites where they are posted or archived.

Table 8-9   Blogs

Engineering & . . . http://blogs.asee.org/engineeringand/ american-waistlines-and-gasoline- consumption-both-expanding Society for engineering Education	Covers developments in engineering as it relates to the world, featuring examples of how engineering intersects with the economy, society, education, and national interests.
Engineering Rapleaf http://blog.rapleaf.com/dev	"... we deal with a lot of engineering issues and obstacles. We cheer amongst ourselves when we come up with a great solution."
CR4 http://cr4.globalspec.com	The Engineers Place for News and Discussion —civil engineering, electrical engineering, engineering, mechanical engineering topics.
UrbanWorkbench http://urbanworkbench.com	Represents the intersection between urban planning, design, and civil engineering.
The Art of Engineering http://blog.engineersimplicity.com	Duncan Drennan puts "down all my thoughts on engineering, business, and creating a better world."
We Know Engineers http:/weknowengineers.blogspot.com	Explores how to manage people; provided by an executive coach.
Engineering Ethics Blog http://engineeringethicsblog.blogspot.com	Comments on current events with an engineering ethics angle.
Mechanical-Engineering Blogs www.mechanicalengineeringblog.com	Provides blog, forums, and website links on mechanical engineering topics.

Table 8-10   Finding Engineering-Related Newsletters

Web search engine such as www.google.com or http://search.yahoo.com	Type the words *newsletter* and *engineering* in the search field. *Note*: Experiment with truncating search terms (use the root of the word): for example, use *engineer**.
Intute: Science, Engineering and Technology www.intute.ac.uk/engineering	Type "journals" in the search box at the Intute Engineering Gateway to get a list of online sources for newsletters and e-journals.
Directory of Open Access Journals: Technology and Engineering www.doaj.org	Select "Technology and Engineering" to retrieve a list of full-text journals in ten subject divisions.
American Society of Civil Engineers www.asce.org	Search for newsletters and journals here.
American Society of Mechanical Engineers www.asme.org	Search for newsletters and journals here.
American Academy of Environmental Engineers www.aaee.net	Search for newsletters and journals here.
Institute of Industrial Engineers www.iienet2.org/Default.aspx	Search for newsletters and journals here.
Society of Automotive Engineers www.sae.org/servlets/index	Search for newsletters and journals here.
World Coal Society: Ecoal www.worldcoal.org	Search for newsletters and journals here.

Why Use Electronic Newsletters.   Electronic journals and newsletters provide technical and practical information of interest to professionals, notices of conferences, current awareness topics, and ads for related services and products. Some of the sites exist primarily to sell products or services or to promote industry or special interests.

How to Find Engineering-Related Electronic Newsletters and Journals.   Many large libraries of institutions offering degrees in engineering maintain lists of newsletters, e-journals, and websites for various disciplines. Table 8-10 provides some strategies for finding engineering-related newsletters.

INTERNET SEARCH TOOLS

Almost all of the access and search techniques previously discussed are integrated and consolidated by the Web. Not only can you search, but you can view, and view not just text but also view graphics and videos and hear audio.

ENGINEERING RESOURCES ON THE WEB

With the Web, you are still limited to (or endangered by) whatever people feel like making available on it. Plenty of engineering resources exist on the Web as the following discussion will show, but it's not like an organized library. Still, there are fascinating resources on the Web, which the list in Table 8-11 helps you access.

Table 8-11   Engineering Resources on the World Wide Web

Infomine: Scholarly Internet Resource Collections http://infomine.ucr.edu	Select Physical Sciences, Engineering, CS, Math and then type your specific topic as the search term. The search engine supports Boolean, truncation, and advanced search techniques.
Intute: Science, Engineering and Technology www.intute.ac.uk/engineering	Part of the larger Intute site, "the Intute Engineering Gateway provides free access to high quality resources on the Internet."
eFunda: Ultimate Online Reference for Engineers www.efunda.com/home.cfm	A portal that provides formulas, mathematics, unit conversions, online calculators, processes, design— even online job search and résumé posting.
EngNet Engineering Directory www.engnetglobal.com	Directory with four main divisions—Engineering Categories, Industry Categories, Brandnames, and Companies—and industry news. Search the site by engineering topic, product, company, or brand name.
Science Accelerator www.scienceaccelerator.gov	"[S]earches science, including R&D results, project descriptions, accomplishments, and more, via resources made available by the Office of Scientific and Technical Information (OSTI), U.S. Department of Energy. Science Accelerator was developed and is made available by OSTI as a free public service."
www.science.gov	"[A] gateway to over 50 million pages of authoritative selected science information provided by U.S. government agencies, including research and development results." You can browse topic categories or search by keyword.
Scirus www.scirus.com	Referring to itself as "the search engine for scientists," Scirus enables searches specifically for conferences in your subject area.
National Academies Press www.nap.edu	Free online access to the full text of a small number of recently published. Scroll to "Engineering and Technology" to find engineering books, or enter a keyword search.
The Online Books Page http://onlinebooks.library .upenn.edu/search.html *or* http://onlinebooks.library .upenn.edu/subjects.html	Provided by the University of Pennsylvania, this resource enables you to find books using its Library of Congress call number and to read the full text online.

Table 8-12   Web Tools for Finding Web Search Tools

Search Engine Showdown http://searchengineshowdown.com	Provides a search engine chart with comparison of features, provides updates on changes in search engines, and reviews engines.
Best Search Engine Quick Guide www.infopeople.org/search/guide.html	Another good starting point for finding the appropriate search engine for your needs.
Best Subject Directories to Use www.lib.berkeley.edu/TeachingLib/ Guides/Internet/SubjDirectories.html	Describes several general subject directories and gives tips on finding more specialized ones. This guide is part of the UC Berkeley Teaching Library Internet Workshops series.
Types of Search Tools www.lib.berkeley.edu/TeachingLib/ Guides/Internet/FindInfo.html	Berkeley Library tutorial covers search engines, directories, the invisible Web, and search techniques, with suggestions on how to decide which to use for your particular needs.
Other Internet Search Tools http://notess.com/search/others	Features specialized search tools such as Email List Directories, tools for searching the "invisible Web," blogs, and free online reference tools.

WORLD WIDE WEB SEARCH TOOLS

How can you survey the entire World Wide Web for engineering resources related to your topic? Use the major search engines already discussed and the subject specific search tools. Table 8-12 lists some websites that can help you find additional search tools.

EXERCISES

If you are not familiar with library-based information sources, find a technical topic that is of interest to you and look for information related to it in as many of the following sources as you can.

1. Check the catalog at your library and WorldCat. For the three most useful-looking books related to your topic, make a bibliographic entry using the format shown in the documentation section of Chapter 11.

2. Using *Ulrich's International Periodicals Directory* to find three useful-looking journals related to your topic, make a bibliographic entry for each one.

3. Using one of the periodical indexes discussed in this chapter, find three useful-looking articles related to your topic in technical journals. Make a bibliographic entry for each one, again using the format shown in Chapter 11.

4. Using NTIS or some similar resource mentioned in the technical reports section of this chapter, find three technical reports related to your topic, and make a bibliographic entry for each.

5. Using one of the patent indexing resources discussed in this chapter, find at least one patent related to your topic, and make a copy of the record that is displayed.

6. Using one of the catalogs described in the product literature section of this chapter, find at least one company involved with products or services related to your topic, and make a bibliographic entry for it.

7. Find one electronic mailing list, one blog, and one website related to your topic. Use the search tools available on the Internet and the World Wide Web to assist you in these searches.

BIBLIOGRAPHY

Berson, Bernard R., and Douglas E. Benner. *Career Success in Engineering: A Guide for Students and New Professionals*. Chicago: Kaplan AEC Education, 2007.

Bird, Linda. *The Complete Guide to Using and Understanding the Internet*. New York: Prentice-Hall, 2003.

Doherty, Paul. *Cyberplaces: The Internet Guide for Architects, Engineers, Contractors and Facility Managers*, 2nd ed. Kingston, MA: R.S. Means, 2000.

Hock, Randolph. *The Extreme Searcher's Internet Handbook: A Guide for the Serious Searcher*. 2nd ed. Medford, NJ: Information Today, 2007.

Lo, Jack and David Pressman. *How to Make Patent Drawings Yourself: Prepare Formal Drawings Required by the U.S. Patent Office*. Berkeley, CA: Nolo Press, 2002.

Lord, Charles R. *Guide to Information Sources in Engineering*. Englewood, CO: Libraries Unlimited, 2000.

MacLeod, Roderick A., and Jim Corlett. *Information Sources in Engineering*, 4th ed. München, Germany: K.G. Saur, 2005.

Neville, Tina M., Deborah B. Henry, and Bruce Neville. *Handbook of Library Research in Science, Technology, and Engineering*. Lanham, MD: Scarecrow Press, 2002.

Osif, Bonnie A. *Using the Engineering Literature*. London, New York: Routledge, 2006.

Pressman, David. *Patent It Yourself*, 13th ed. Berkeley, CA: Nolo Press, 2008.

9

ENGINEERING YOUR SPEAKING

The podium or lectern can be a lonely and intimidating place.... Despite the fact that they can help make or break a person's career, oral presentations often turn out to be boring, confusing, unconvincing, or too long. Many are delivered ineptly, with the presenter losing her or his place, fumbling through notes, apologizing for forgetting something, or generally seeming disorganized and unprofessional.

John Lannon and Laura Gurak, *Technical Communication*, 12th ed. (New York: Pearson Longman, 2010).

I was gratified to be able to answer promptly, and I did. I said I didn't know.

Mark Twain, 1835–1910.

Engineers are often called on to speak formally, and many engineers find they have to speak a lot. Whether you give an impromptu 5-minute briefing or a professional 1-hour presentation at a technical seminar (or something in between), you should see your talk as a great opportunity to share information and to show that you know how to communicate. Few of us are naturally gifted with such capabilities, and some of us are almost petrified at the thought of talking before a group, but the skills possessed by good speakers can be learned. The principles discussed in this chapter will enable you to become a confident speaker people will listen to, because you transfer information efficiently and effectively—that is, with a minimum of noise.

PREPARING THE PRESENTATION

Developing a worthwhile presentation is like developing a product: Research and planning are crucial in the early stages. We all know what it's like to have to come up with a spontaneous briefing or unexpected oral report, but fortunately we usually have some lead time before we talk. Using that time to work through the procedures that follow will help you design a successful presentation.

> **Portable sign-language translator**
>
> A group of engineering-technology and industrial-design students from the University of Houston have created MyVoice. The device uses a camera to "read" the hand gestures of a deaf person and then audibly "say" the message to the hearing person via a soundboard and speaker.
>
> For details, see the Preface for the URL.

ANALYZE YOUR AUDIENCE

Much of what was said at the beginning of Chapter 3 about focusing on your reader and purpose *before* writing should also be applied to preparing for an oral presentation. We've all been bored by talks that were over our heads, too simplistic, or unrelated to our interests. Don Christiansen, a former editor and publisher of *IEEE Spectrum*, humorously recounts one of his early experiences:

> *As a young engineer, I was invited to address an IEEE Section meeting. My subject was an unusual stereophonic/quadraphonic audio system developed at CBS Laboratories. This technical presentation may have been my first before a large engineering audience. I worried at the prospect. I prepared and projected a number of slides containing a bunch of mathematics that no one could follow during a brief exposure. After all, I had sat through many conference papers that were ritually peppered with unintelligible (at least to me) equations. I had responded in kind, despite my audience having many spouses present—most of whom hadn't a clue what their mates did for a living. I was grateful to the wives, who did not boo or stamp their feet, but discreetly nodded off.*
>
> Donald Christiansen, "Engineers Can't Write? Sez Who!,"
> *IEEE-USA News & Views*, June 2003, p. 4.

To make sure you don't do the same thing to others, ask yourself the questions listed in Table 9-1 when beginning to prepare your talk, and make sure you have as clear an idea of the answers as possible.

Table 9-1   Some Questions to Ask About Your Listeners Before You Talk

- Who will the key individuals in my audience be?
- What needs or concerns do they have regarding my topic?
- What are *my* objectives for this talk?
- How knowledgeable are my listeners about my subject?
- How can I get their attention and interest right away—and keep it?
- What are their attitudes likely to be regarding what I have to say?
- Do I need to work on changing their attitudes, and if so, what is the best way to go about it?
- What benefits are they going to get from listening to me?
- What kinds of questions are they likely to ask?
- What kind of feedback do I want?

DECIDE ON YOUR PRIMARY PURPOSE

Your purpose in talking is intimately related to the makeup of your listeners and the reason they are sitting in front of you. Are they there for instruction, information, insight, to be persuaded, or what? What action or change do you feel they need to undertake? Knowing exactly what kind of assignment you have will also determine your foremost purpose. Engineering presentations can take many forms, as Figure 9-1 indicates, each with a specific purpose and organizational requirements.

Figure 9-1   Just a few of the many kinds of presentations engineers find themselves giving.

Make sure you know what you are getting into, what is expected of you, who your audience is going to be, and what you want to accomplish by talking to them. Decide exactly what you want your listeners to take away from your talk. Then you will be on solid ground while preparing the remaining features of your presentation.

DETERMINE YOUR TIME FRAME

It has been said that no speech is ever too long for the speaker or too short for the listener. The cardinal rule here is *never* to speak longer than you are supposed to. To avoid annoying busy people or offending speakers who come after you, check how much time you have been allotted. Knowing your time limit will also help you decide how much detail you can go into, how much time you should allow for questions or discussion, and how much time you can spend on an introduction and conclusion or recommendations if you have some.

As Figure 9-2 illustrates, how deeply you go into different aspects of a typical engineering topic is related to how much time you have to speak. The tops of the pyramids in the figure represent the least you could say on a topic—perhaps a single sentence—while the true base of the pyramid (unseen in the illustration) represents everything that could possibly be said. This is perhaps why we almost always impose time limits on speakers; otherwise they might go on forever!

It's been claimed that any subject can be covered in virtually any amount of time. A speaker could compress the creation of the universe into three or four sentences—or fewer—if necessary. In the same way, an expert could talk for hours (probably to a rather small audience) on the mating habits of the Gambian giant pouched rat. Whatever you do, don't decide that since you have a lot to say in a short time, you should speak as fast as possible while rapidly clicking slides on the screen. You will just as rapidly lose your listeners.

IDENTIFY YOUR KEY POINTS

As indicated above, oral presentations normally don't permit us to give all the miniscule details about every aspect of our topic. So don't expect to say everything that could possibly be said on your subject. With a sharp awareness of your main purpose and time frame, decide *what the most important points are* that you want to get across to your audience and how you want to develop those points in the time you've got.

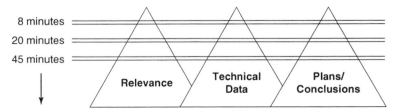

Figure 9-2 No matter how little time you have to talk, you can say something worthwhile in an engineering presentation. You just have to correlate how deeply you delve with the time that you have.

You may need to be quite mathematical about this. If you have 20 minutes to make four important points, you may subtract the time you want for an introduction and conclusion and divide what is left by four, thus leaving from three to four minutes for each point. In some presentations you may not want to give equal time to each point. For instance, if you have to discuss five reasons why a project should be canceled or why your company should invest in new equipment, you might determine which points will meet the most resistance from your audience. Then aim to spend more time explaining those points while giving briefer treatment to the less controversial ones.

CHOOSE AN ORGANIZATIONAL PLAN

Your subject and purpose, and to some extent the time you have, will help determine how to organize your material. Most presentations can roughly be broken down into an introduction, the main points, a conclusion, and a question-and-answer period. Table 9-2 is a list of ways to organize your central material. Some presentations will call for combining some of these organizational plans, of course.

Table 9-2   Ways to Organize Material for an Effective Oral Presentation

Time Sequence	Describe progress on a project or steps in a procedure. Relate decisions leading up to an action or occurrences that led to a problem.
Spatial Sequence	Describe equipment or a physical area such as a test site or plant facilities. Outline a physical process.
Decreasing Importance	Give your most important points first down to the least important: relating six ways to improve or prevent a situation.
Increasing Importance	Work up to your most important point: some minor reasons for an action, change, or decision, followed by the major reasons.
General to Specific	Present a general point followed by specific examples: "We've got to improve production," followed by concrete ways to do so.
Specific to General	Be persuasive: citing examples of personal injury to lead to the point that more stringent safety regulations are needed at your plant.
Comparative	Compare and contrast equipment, approaches, or ideas on such aspects as costs, durability, reliability, ease of operation.
Familiarity	Begin with the familiar first, leading your audience into an understanding of the unfamiliar: talking about corporate needs or problems.
Difficulty	Present data in order of the easiest first and progressing to the hardest, as in a training session or tutorial.
Controversiality	Begin with least controversial points in order to be diplomatic about sensitive issues: why changes should be made in a project in which people have some ego investment.

PREPARE AN OUTLINE AND NOTES

Writing an outline and notes helps you clarify in your own mind how best to present your material. They also give you a means of deciding how much time to allot to each point, and they will be helpful when you rehearse the presentation. However, extensively relying on an outline or notes during the actual presentation can be dangerous, because you will give the impression that you don't know your topic thoroughly.

Your notes and outline may range from a few hastily scribbled ideas—if that much—jotted down a few moments before an unexpected briefing, to a complete manuscript of every word you intend to say. Reading a word-for-word written version of your talk in front of an actual audience is **NOT** a good idea, however, unless you feel very insecure or are giving a highly technical conference paper calling for extreme precision and accuracy. Even then, few people want to sit while a paper is read to them; after all, they could read it themselves in the comfort of their own homes or offices.

While preparing your presentation, determine which prompts will best keep you on track when giving it. The main cues engineers tend to use are the following:

- An **outline** of the complete talk, with key ideas highlighted or in large print to be quickly glanced at if necessary as the presentation goes along (see Figure 9-3).
- **Note cards** numbered in the order they will be used, with key ideas and facts clearly written on them. If these are relied on too much, however, you will appear unsure of your material.
- **Visual aids**, such as transparencies or slides. If you are really on top of your topic, your visuals themselves will be all you need to keep on track. They may,

---

I.	**Future Needs**
	a. Sharper Images
	b. Cheaper Cyclotrons
II.	**Safety**
III.	**Cost** of Current Technology
	a. Combination with MRI
IV.	Is It **Profitable?**
	a. To **Market Now?**
	b. For Research?
V.	**Company's Interest**
VI.	**Role** of the Company
VII.	**Recommendations**

Figure 9-3 Part of an outline to aid the speaker during an oral presentation. Key points are highlighted in bold.

in fact, be the outline of your talk. If you wish, you can make printed copies of them for your own use, with notes written to yourself that can be quickly referred to if needed.

- **A backup plan** in case something goes wrong with the equipment you're using. If you have attended many meetings or seminars, you know this still can happen, so you may want to have some hard copies of your overheads to hand out in case of a system failure.

CREATE SUPPORTING GRAPHICS

Since we live in an increasingly visual age, and since people remember information better when they both hear it and see it, most effective engineering speakers support their talks with illustrations of some kind. Graphics also work to your advantage, since preparing them forces you to organize and rehearse your presentation and possibly discover weak spots that need attention. Showing them will save you time during the presentation since you won't have to write the information down on a board or flip chart. They can also serve as cues for you, reminding you of what you want to cover and the order in which you want to cover it. You should at a minimum plan to use visuals wherever you feel they will:

- Simplify a point.
- Clarify a point.
- Stress a point.
- Show critical relationships between ideas or facts.

Channels for Graphic Support.   For engineers, one of the most common means of showing graphics has traditionally been the overhead projector with *transparencies* (also known as foils, overheads, visuals, or view graphs) that you prepare in advance. Transparencies have the advantage of letting you add information on them with a wax marker while showing them, or even using additional ones as overlays. If you have several transparencies, remember they can be slippery. It's embarrassing to see them slide off the table and across the floor, so you might want to consider a matting of some sort for each one. *Flip charts* are useful for ongoing illustrations or emphasis during your talk, as is a *chalkboard*. Slides shown from a *slide projector* can be effective also, especially if you can do a professional job with color and art work—but this can be time-consuming and is now considered rather old-fashioned.

The most popular way to display visuals nowadays is from a *laptop computer* connected to a projector, showing the audience what you have created on a graphics program such as PowerPoint, OpenOffice, or Harvard Graphics. Presentation software such as these allow you to progress through your talk by calling up graphics through your keyboard or a wireless mouse. Such programs also allow you to add numerous features to your presentation, such as sound, zooming text, and colorful templates. The danger here is the temptation to get too fancy and to try to dazzle your audience with your artistic skills rather than by presenting clear, visually accessible information.

Furthermore, the most impressive visuals you can make will not lessen your need to speak clearly, effectively, and enthusiastically.

If you are talking about a specific piece of equipment that you can bring into the room with you, do so—as long as your audience is small and close enough to be able to see it. If you hand it around, remember to get it back. This might seem obvious, but in the afterglow of a good presentation, with people crowding around you with praise or questions, it's easy to forget to retrieve your widget.

Designing Your Graphics.   Any good graphics program will give you everything you need to create graphs, pie charts, bar charts, flow charts, and any other examples. Nowadays with only a few hours' training, any engineer can produce almost any kind of graphic with programs like PowerPoint. Some specialized programs allow you to create excellent illustrations of equations, electrical circuits, and other technical data. With the growing use of scanners, you can now copy and present professional illustrations or photographs of just about anything. If you show scanned material in your presentation, be aware of any copyright restrictions that might apply, and give credit (usually at the bottom of your slide in small print) to the source of any such material. Chapter 11 of this book deals with the question of citing your sources.

A WORD OF CAUTION

Although presentation software, especially PowerPoint, is widely used practically everywhere nowadays and can produce stunning graphics, it is not universally admired or thought to be the solution to every presentation task. Professional communication experts often warn of the dangers of visually presenting information as no more than a series of bulleted points that oversimplify or dumb down our material. Other critics warn of incompatible color combinations being used or of clever artwork on the screen that overwhelms the subject. These possible negatives are constantly debated and analyzed and if you want to become familiar with them, enter "Drawbacks of PowerPoint" or "Dangers of PowerPoint"—or some similar phrase—in your search engine.

MAKE YOUR INFORMATION ACCESSIBLE

As we implied earlier, the most dazzling transparency or slide will impress no one if the information it contains is not easily accessible. In fact, anything you put on the screen that cannot readily be grasped by your audience—because it's either (a) *too complex* or (b) *too small*—is worthless. This point is particularly worth heeding if only because we see it so often ignored by engineering speakers.

  (a) **Too complex.** Don't let your visuals suffer from *information overload*. Each should be as simple as possible, portraying the bare facts—you can always elaborate verbally. Even quite technical material can be reduced to manageable concepts on a screen. For example, something as complex as an electronics circuit can be broken up into constituent parts after a simplified overview has been given, as shown in Figures 9-4 to 9-6.

(b) Too small. If your visuals consist of lists or other written information, *make the words easy to read* (see Figure 9-7). This means using at least a 24-point font size, preferably boldface. It's best to have no more than eight lines of print on a slide or transparency, and better not to use all capital letters since this makes for harder reading. A page of text or an illustration photocopied from a book or journal rarely makes a good overhead.

When you present written information on the screen, don't crowd it. Provide ample margins and plenty of white space between and around the lines. You might want to use bullets, checks, or other marks to emphasize points, but resist the temptation to go overboard with the variety of fonts and clip art now available. Don't let your artwork overwhelm the information on the screen.

Brainwave communication

In 2012, tech startup Neurovigil and Stephen Hawking have been testing the potential of its iBrain device to allow the astrophysicist to communicate through brainwaves alone—a neural, not an oral, presentation.

For details, see the Preface for the URL.

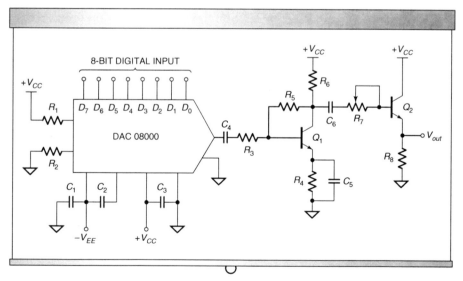

Figure 9-4   An example of an overcrowded transparency. Far too much information is thrust upon the audience here. One way to make this material more accessible would be to reduce the circuit first to a block diagram, as shown in Figure 9-5, and then, if more detail is needed, to expand the drawing one block at a time on separate visuals, as in Figure 9-6.

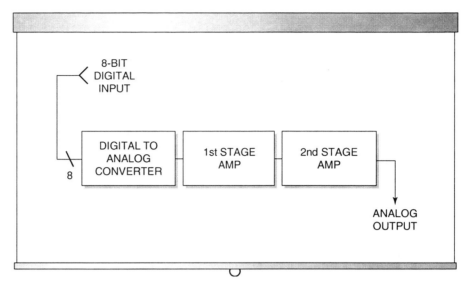

Figure 9-5   Simplified block diagram of the circuit in Figure 9-4.

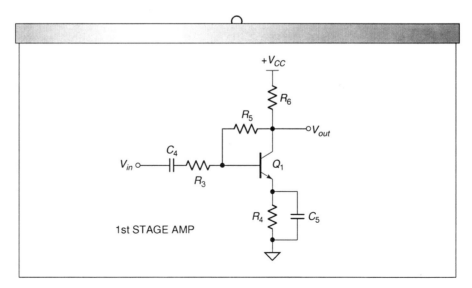

Figure 9-6   The center block from the diagram expanded to show part of the original circuit.

PREPARE HANDOUTS

Think carefully about whether you want to provide handouts, and if so, what kind and when you will hand them out. Many speakers wisely avoid handouts altogether, as they feel they distract their audience's attention. On the other hand, some speakers

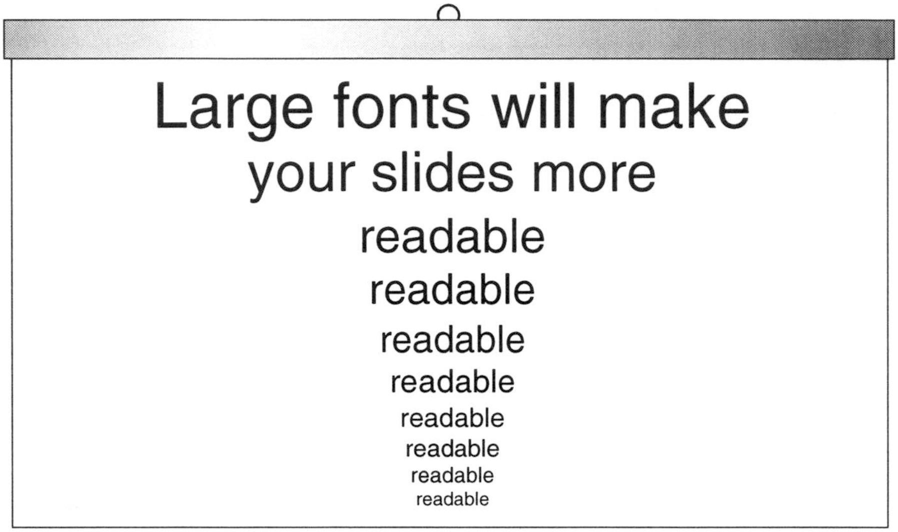

Figure 9-7   Use at least 24-point print on your overheads if you want your audience to read them.

pass out copies of their overheads or slides, often reduced in size, so listeners can make notes on them. Distributing an outline of your presentation may be a good idea, especially if the topic is detailed and covers a lot of material, but the choice should be yours. Sometimes you might need to provide supporting evidence for your talk, such as samples, brochures, or other data. Plenty of successful speakers, however, expect their listeners to focus solely on the presentation itself and to take their own notes as they wish. The dilemma with handouts is *when to pass them out*. If it's at the wrong time, they will almost certainly distract from your talk, since people tend to look at what is given them right away and may ignore you or only partially listen. You need to decide beforehand when the best time to distribute material will be, and how and when you will refer to it, so that it adds to your listeners' concentration and understanding rather than takes away from it.

PREPARE YOUR INTRODUCTION

When thinking of how to begin your talk, remember that

1. Your audience may be asking themselves, at least subconsciously, "Why do I need to hear this?" or "Why should I be here right now?"
2. Your audience has a limited attention span.

To help solve both problems, design your introduction to *let your audience know right away what your topic is and of what benefit it is to them.* In essence, tell them *why* they should listen to you. Then provide a sense of direction by giving an overview of where you're going in the presentation and what you plan to cover. Let your audience know how long you intend to speak if it's not already known. Many speakers lose their audience right away because they fail to follow these procedures at the outset of their presentation.

PREPARE YOUR CONCLUSION

Design the end of your presentation to focus the audience's attention solely on essentials. Depending on the type of presentation you give you will be able to reinforce your message by, for example:

- Summarizing what you have discussed.
- Stressing your central idea once more.
- Reviewing your key points.
- Restating your recommendations or decisions.

An appropriate final summing-up slide can be a great help here. Above all, give your audience a lasting impression of what you want them to take away from your talk, such as the feeling that you have solved a problem or concern or have provided new insights. Don't suddenly stop talking at this point, however. Make a note to close gracefully with something like "And this concludes my presentation. Thank you for your attention. Are there any questions?"

GET READY FOR QUESTIONS

If there is going to be a question period after your talk, spend some time during the preparation stage to anticipate and get ready for them. Put yourself in the place of your listeners: Are they likely to find any part of your talk especially difficult, detailed, or controversial? Are they likely to hold any opposing viewpoints? Are there areas you may not be able to go into as thoroughly as you would like, due to time restraints, and which might therefore generate questions? What could be the "worst" question asked? Also, can you think of diplomatic ways to encourage questions from people who are reluctant to ask? (Sometimes a friendly smile or "I'd *love* to have some questions" is all that's needed.)

When questions come, it's often a good idea to repeat them aloud in some form before answering. The repetition is useful for people who didn't hear the question clearly in the first place and the delay might give you time to gather your thoughts.

PRACTICE, PRACTICE, PRACTICE

Keep in mind these letters: **PPPPP** (Plentiful Practice Prevents Painfully Poor Presentations). Some speakers go over their material—outline, notes, visuals—up to seven times before presenting it; for others this would be overkill. Most speakers do rehearse at least twice, however, if their talk is of any significance or if they feel unsure of their material.

Depending on the importance of your talk, you may decide to have at least one dress rehearsal if you can, preferably in the room where you'll be presenting. This will let you get familiar with the room and any equipment to be used. An audio- or videotape of this rehearsal, if possible, would enable you to self-critique your performance. On the other hand, you might want to find a trial audience to listen to your first run and give you some feedback. Friends, colleagues, a spouse, or even yourself in the bathroom mirror, can be good audiences to practice on.

Perhaps the most valuable outcome of careful practice is the self-confidence you gain. One antidote to nervousness about speaking in front of a group is to be able to walk into that room knowing you're completely in control of your subject and ready to present it in an effective manner—confidence you can only gain by first practicing as much as possible.

DELIVERING THE PRESENTATION

When was the last time that you sat through two and a half hours of a scientific presentation and wished that it would go longer?

Michel Alley, *The Craft of Scientific Presentations* (New York: Springer-Verlag, 2007), p. vii.

All your preparation efforts are aimed at one goal: to give an effective, noise-free presentation that will produce the desired results. By the time you stand in front of your audience, you should have fixed many of the potential glitches that can surface in oral presentations. Knowing your subject and audience makeup will have helped you determine the information you need and how you need to communicate it.

Most engineers can prepare a presentation well enough given a little awareness, analysis, and preparation time, yet the sad fact remains that plenty of lackluster and somewhat boring presentations occur every day in business and industry. As with a written report, such presentations can be greatly improved by the elimination of noise.

AVOIDING NOISE IN ENGINEERING PRESENTATIONS

In an oral presentation, noise can be defined as anything that prevents the message from effectively getting into the minds of the audience. Following are some causes of noise that frequently occur in engineering (and other) presentations.

1. **Speaking Too Softly.** Avoid the tendency to speak too softly. Try to project your voice relative to the room and audience size. If some listeners can't hear you, you're wasting their time.

2. **Speaking Too Slowly or Rapidly.** Avoid a slow, labored pace with too many pauses. Also avoid speaking too rapidly. Aim for a speed slightly slower than

normal conversation. Remember that pausing and deliberately slowing down once in a while can help you stress important points.

3. **Speaking Monotonously.** Ever heard the joke about the dull college professor who dreamed he was giving a lecture only to wake up and find he was? How you talk often makes a bigger impression than flashy visuals and what you say combined. You could be explaining the never-before-revealed secrets of time travel and yet find few paying attention if you sound bored to death. Hypnotic monotony can be avoided by varying your pace and your pitch—by speaking the way most people do in lively and energetic conversation. **ENTHUSIASM** on your part will encourage your audience to listen to you.

4. **Using Verbal Fillers.** When a speaker needs to pause or is uncertain of what to say next, irritating and empty catchwords or phrases like *uh, umm, basically*, and *yuh no* sometimes take over. Don't distract your audience with a high UPM (umms per minute) rate. Try to avoid this kind of noise in your presentations. There's nothing wrong with being silent for a few moments while gathering your thoughts.

5. **Becoming a Statue, Pendulum, or Traveler.** Avoid the tendency to freeze up physically when in front of a group. Also avoid the tendency to sway back and forth without moving your feet as well as frantic pacing and hand movement. Both are distracting and do not add liveliness to the presentation. Try for a natural stance and movements when in front of your listeners, with some foot movement but not enough to wear out your audience as they follow you back and forth across the room.

6. **Blocking the Screen.** Avoid standing directly in front of the screen and staying there throughout your talk (see Figure 9-8). It's just as bad to stand partially off to the side yet still block the screen for those sitting at the sides. If you can't avoid blocking the screen, move around enough during your presentation so that you don't block anyone's view continuously.

7. **Reading from the Screen or from Notes.** Keep to a minimum any reading of your slides or notes during a presentation. Straight reading can easily become monotonous. Also, be wary of dimming the lights to make your visuals easier to read. Low lights can make people drowsy and can hide facial expressions and the eye contact you need to have with your listeners. Additionally, reading from notes or frequently referring to them will straightaway give your audience the impression that you don't know your subject as well as you should.

STRENGTHENING YOUR PRESENTATIONS

Use an Informal Style.   When making an engineering presentation, you're not delivering a sermon (usually) or pronouncing on a profound legal intricacy. Generally, the best style is an informal one, paralleling as closely as possible the normal conversational mode of everyday life. It's quite all right to use contractions (*it's, don't, couldn't,* etc.), even if you avoid them in formal writing. Using pronouns such as *you, your, I,* and *we*

Figure 9-8   You invest a lot of planning and work into your visuals, so don't create noise by standing between them and your audience.

will help you relate to your audience's interests and needs and will indicate you are interested in them as people rather than as an impersonal mass. Avoid any technical jargon not readily understood by your listeners.

Make Clear Transitions.   Even if you have a well-organized talk full of important details, be sure to show the connections between your ideas. Your visuals will assist you, of course, but make your presentation as user friendly as possible. Keep your listeners in the picture by emphasizing connections and transitions in your thinking by using simple words and phrases like these:

• First	• On the other hand
• Next	• As you can see
• To begin with	• For example
• Initially	• Also
• Furthermore	• Finally
• Consequently	• In conclusion
• As a result	• To sum up

When you overlook such transitions in a written report, your readers can at least go back over the material and figure out the connections for themselves, exasperating as that might be. Someone listening to a talk has no such opportunity.

The effective presentation resource at Penn State (www.engr.psu.edu/speaking /SPEECH.html) provides an excellent example. Matt Chang in his presentation entitled

"Graphene: The Answer to Moore's Law?" uses transitions phrased as indirect questions:

> *You might be wondering at this point, why is it a problem if my computer can't do over a billion computations every second?*
> *You might be wondering, how can we combat a problem like this?*
> *You have to be thinking, how can we tackle this problem if we are already down to the atomic scale?*
> *But you might be thinking, "One atom is weak and how can we construct something that is one atom thick? Won't it just break?"*

These questions introduce each new section of the presentations, spaced about five minutes apart.

Repeat Key Points.   No matter how brief the presentation, you're going to have at least one main point you want your audience to go away with. Don't be afraid to repeat yourself—your listeners need to know the most important aspects of your subject. There is a lot to be said for that old piece of advice, "*Tell your listeners what you're going to tell them, then tell them, and finally tell them what you have told them.*" It's essential to repeat your key points in a concluding summary.

Use a Pointer.   A pointer is the best way to focus your audience's attention on your key points while you explain what they're looking at. Avoid *laser pens* because they are distracting, easily misused, and potentially harmful to people's eyes.

Note   Some people are unable to see a small laser dot on a screen.

A *straight metal or wooden stick* pointer is always available, but you have to stand fairly close to the screen to use it, possibly blocking the view for some people. The *retractable stick pointer* has the same potential drawbacks. If you use a stick pointer, hold it with the arm closest to the screen so you don't have to turn away from the audience every time you point (see Figures 9-9 and 9-10). Some speakers have a nervous tendency to open and close this kind of pointer repeatedly, or to even scratch themselves with it. When not actually using it, keep your pointer firmly clasped in one or both hands and resist the temptation to conduct an imagined symphony with it.

Maintain Eye Contact.   You increase your credibility a great deal by looking at your audience as you talk. While avoiding eye contact could give the impression that you're shifty or unprepared, looking at your audience helps establish rapport with each member in a small group and creates a sense of intimacy with a larger group.

As you progress in your talk, try to hold visual contact with each person for a few seconds and move on to someone else. Looking at individuals also enables you to pick up feedback on how they are receiving your message; puzzled looks or frantic note-taking, for example, might show that you need to go back over something or slow down.

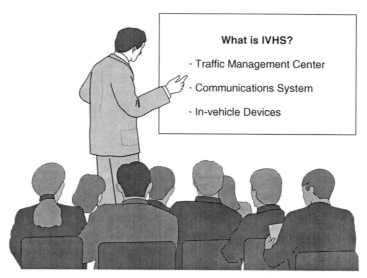

Figure 9-9   Using the arm farthest from the screen to point is a problem: It pulls you from eye contact with your audience.

Figure 9-10   Using the arm closest to the screen still allows you to talk while facing your audience.

Be Ready for Unexpected Questions.   You can prepare all day for questions and still land at least one question you never dreamed of. Try not to appear surprised or defensive when this happens—you've prepared a lot and know your subject well. Two strategies for tackling unexpected questions are to:

- Simply say you don't know. People will respect you for being honest, and you can still offer to supply possible sources for the information or research the answer later.
- Offer to talk with the questioner after your presentation. This may be the best answer if (a) the question is too involved for the discussion you are in, (b) the question is not really related to your topic, or (c) the question is hostile and you don't want to get into an argument. Rarely will anyone seek you out afterwards unless he or she is genuinely interested in information you may have.

Accept Your Nervousness.   If you're just starting to give presentations, unfortunately the best cure for nervousness is experience. Until then, accept your nervousness as perfectly normal. We all suffer from it (although our nervousness is often much less noticeable to others than we might think). Learn to use this anxiety as a positive energizing power that helps you to be more alert and lively. If you have stage fright, consider the following tips:

- Enter the presentation room knowing you've worked hard on your presentation and have practiced delivering it. In others words, give yourself as much reason as possible *beforehand* to be confident of your knowledge and ability. Then try to concentrate on your topic rather than on yourself. Appropriate gestures and facial expressions will often occur naturally in a well-prepared speech.
- Take some deep breaths before entering the room. Even a short walk around the building or a few simple physical exercises may help relieve anxiety. However, these activities can never substitute for the self-assurance that comes with really knowing your material.

PRESENTING AS A TEAM

Since engineers frequently collaborate on a project, compile a proposal, or report on a new product, you are likely to be involved in team presentations. These allow individuals to speak in turn on a topic, each with his or her own part. Group presentations also permit a specific aspect of a complex subject to be presented by the individual who worked on it rather than by a team spokesperson. This kind of presentation is additionally effective because:

- Teamwork reduces everyone's preparation workload.
- Longer presentations are possible without exhausting one person.
- Speakers can enjoy team support during the presentation.
- The variety of speakers helps hold the audience's attention.
- Each topic can be explained (and questions answered) expertly.

PREPARING FOR A TEAM PRESENTATION

Whether team or solo, your first step is to analyze your audience and purpose. Decide on the main points to be stressed, their order, and who will cover which topic. The team leader should clearly partition the topics and make sure that each speaker sticks to the assigned topic and doesn't cover any other speaker's material.

It's essential to allocate time to each speaker early so that everyone can prepare accordingly and the presentation can conform to any required time limits. Decide beforehand whether questions from the audience should wait until after the entire presentation or should follow each speaker.

SHARING THE PRESENTATION

Assigning different parts of the presentation to alternate speakers prevents a long presentation from becoming monotonous, arouses audience interest, and provides clear structure to the presentation. Plus, each speaker gets to have a breather.

To ensure that your group presentation flows well, pay attention to how you are going to move from one speaker to the next. A simple lead-in like "... and now Eva is going to cover the financial aspects of the project," might be all you need. If there are just two of you, break down the topic so you can alternate with each other several times.

> ### Speech glove communication
>
> AnthroTronix, an engineering research and development firm, has developed AcceleGlove, an open-source "data glove" that, worn by a deaf person, translates that person's gestures into written text or spoken words. The device was originally conceptualized in the doctoral dissertation of José-Luis Hernández-Rebollar.
>
> For details, see the Preface for the URL.

MAKING A DRY RUN

Groups, like individual speakers, need to put aside plenty of time for practice. Rehearsals will uncover any information gaps, neglected subtopics, or transition problems. As an individual team member, make sure that you can conform to time limits: There is no better way to make enemies than to dominate the presentation and speak longer than you should, shrinking fellow speakers' time.

CHECKLIST FOR ORAL PRESENTATIONS

The items in Figure 9-11 can be used to evaluate a presentation. You can also use the list *before* you give your talk. Put yourself in the place of your audience and try to get a sense of how they would "grade" you as they listen to your presentation. You might even get a friend to check you out on each item during a dry run.

Introduction	Visual Aids
☐ Creates favorable atmosphere	☐ Are clear and easy to read
☐ Creates an appropriate pace	☐ Look professional
☐ Hooks listeners' attention	☐ Avoid information overload
☐ Relates subject to listeners	☐ Clearly support related ideas
☐ Presents clear central idea	☐ Are enough
Body	**Delivery: Sound**
☐ Reveals careful audience analysis	☐ Clear volume and pronunciation
☐ Supports central idea	☐ Effective diction
☐ Maintains audience interest	☐ Varied speech patterns
☐ Provides technical accuracy	☐ Absence of *uh-huh, y'know, like, basically*
☐ Organizes details effectively	☐ Enthusiasm
☐ Allocates time carefully	☐ Standard grammar and usage
☐ Provides clear transitions	☐ Good question response
Conclusion	**Delivery: Appearance**
☐ Ties presentation together	☐ Professional posture and appearance
☐ Restates central idea	☐ Appropriate gestures and mannerisms
☐ Proposes action or seeks response	☐ Effective use of pointer
☐ Invites discussion or questions	☐ Consistent eye contact with audience
Attention to Limits	☐ Competent handling of notes and visuals
☐ Too short	
☐ Just right	
☐ Too long	

Figure 9-11   Oral presentation checklist.

LISTENING TO PRESENTATIONS

You will probably listen to more presentations than you give during your engineering career, yet listening is the most neglected of all communication skills. Nothing is more frustrating than working long on a presentation just to have an unresponsive or uninterested audience. To be sure, the responsibility for a successful presentation lies partly with the audience. Here are some ways to be a good listener:

- Maintain natural eye contact with the speaker.
- Show that you are alert, interested, and well-disposed toward the speaker.
- Ignore distractions such as people talking or other external noise.
- Take notes on the speaker's most important points.
- Develop at least one question in your mind, and ask it at the appropriate time.
- Be sure to turn off all cell phones and beepers.

In fact, one of the best ways to be a good listener is to ask questions. A question lets the speaker know you're paying attention and that the presentation has made you think. Even if you understand everything presented, why not ask about something you found particularly interesting or new?

Actively focusing on a speaker and concentrating on what is being said establishes a sense of empathy that leads to more efficient information transfer. Being an active listener is not just a matter of being kind to the speaker; it also rewards you with a more complete appreciation, knowledge, and evaluation of the material presented. You go away a more informed person.

THE IMPORTANCE OF INFORMAL COMMUNICATION

Vital as the ability to give effective presentations is to your career, another aspect of oral communication is seen as trivial, even though it's not. This can simply be considered "small talk," or informal chatting on the job.

At work it might seem like a time waster—you may well be familiar with the figures of Dilbert and Wally, coffee cups in hand, getting nowhere in a hurry. Yet the water fountain, coffee machine, or cubicle area, as well as the elevator or parking lot, can all be places where you can benefit from efficient conversation. Waiting for a meeting to begin or attending a professional conference are also great times for anything from friendly chitchatting to more serious networking. Labtec's Senior Vice President of Technology and Engineering, Todd Yuzuriha, in his book *How to Succeed as an Engineer: A Practical Guide to Enhance Your Career* (Vancouver, BC: J&K Publishing, 1999), puts it this way:

> *Take initiative to build good relationships with your co-workers. As organizations are becoming more interdependent, cooperation and collaboration among co-workers is essential. Be a team player. Look out for the needs of others as well as your own. Taking the time to build good relationships with your co-workers not only makes your work environment more enjoyable, it can help you and your organization get results (p. 47).*

Small talk is not a waste of time when it's used to personally or professionally connect with people. Some studies have shown that the ability to engage others in conversation can have a good effect on your career. According to Margo Frey, writing in the *Milwaukee Journal Sentinel*, informal small talk can make or break your career, and Debra Fine, a former engineer, has written at least one book on how small talk can help you in your profession. If you need more convincing, plus many helpful tips on how to become adept at effective small talk, look at Fine's book (listed in the Bibliography at the end of this chapter), or do a web search on "professional small talk" or "small talk on the job."

EXERCISES

1. Ask one or two engineers about the oral presentations they regularly make. How long do they talk? What do they talk about? Who is their audience, and how do they meet their needs? Do they use any graphics or handouts? What kinds of feedback do they get? How do they know whether their presentation has been successful?

2. Listen to someone giving an oral presentation and evaluate his or her performance as best you can using the checklist in Figure 9-11. After evaluating the presentation, think of ways you might improve it if you had to give it yourself.

3. Take a written report—your own, if possible—and turn it into an oral presentation. Do you have to leave out some of the material or change its order or emphasis? Is there anything that can be presented graphically? How will you introduce and conclude your presentation? Does the fact that you may be asked questions after the talk cause you to think differently about your material than you did when writing the report?

4. Think of the various people you have listened to, such as teachers, fellow professionals, preachers, or politicians. What are some of the best things you have heard or seen these people do? What are some of the worst? Which were the most effective speakers? Which were the most ineffective? What can you learn from both kinds that will help you improve your own skills in giving oral presentations?

BIBLIOGRAPHY

Alley, Michael. *The Craft of Scientific Presentations: Critical Steps to Succeed and Critical Errors to Avoid.* New York: Springer-Verlag, 2002.

Beebe, Steven A., and Susan J. Beebe. *Public Speaking: An Audience-Centered Approach.* Boston: Allyn & Bacon, 2011.

Beer, David F., ed. *Writing and Speaking in the Technology Professions.* New York: Wiley–IEEE Press, 2003.

Fine, Debra. *The Fine Art of Small Talk: How to Start a Conversation, Keep It Going, Build Networking Skills—and Leave a Positive Impression!* New York: Hyperion, 2005.

Fisher, Donna. *Professional Networking for Dummies.* New York: For Dummies, 2001.

Indiana University. Small talk made easy. http://newsinfo.iu.edu/news/page/normal/2500.html. Accessed July 3, 2012.

Penn State University. Effective Presentations in Engineering and Science: Guidelines and Video Examples. www.engr.psu.edu/speaking/. Accessed July 3, 2012.

Tufte, Edward R. *The Visual Display of Quantitative Information.* Cheshire, CT: Graphics Press, 2001.

10

WRITING TO GET AN ENGINEERING JOB

The résumé is your main vehicle for presenting yourself to a potential employer. The central question to ask in preparing your résumé is, "If you were an employer, would you want to read this résumé?"... Visual impact and appearance are extremely important.

> Raymond Landis, *Studying Engineering: A Road Map to a Rewarding Career*, 2nd ed. (Los Angeles, CA: Discovery Press, 2000), p. 223.

The key to résumé writing excellence is in presenting it the right way. Most people make the error of just listing their experience and qualifications; this ends up being a rather boring document. A good résumé should not only demonstrate your skills and experience, but should also give the reader a good indication of the type of person you are. It needs to have personality.

> Engineers International, "Preparing the CV,"
> www.engineers-international.com/careerscv.html.
> Accessed July 16, 2003.

Two tools commonly used to seek employment are the résumé and the application letter.[1] You send one or both of these to prospective employers when you are in a job search. The combination depends on the potential employer—some request only the résumé, some request only the letter, and some don't indicate. When you're not sure, send both.

---

[1] Our special thanks to Randy Schrecengost, P. E., for his reviews and recommendations on this chapter.

Note   Additional examples are available at the companion website for this book. See the Preface for the web address.

HOW TO WRITE AN ENGINEERING RÉSUMÉ

A résumé is a summary of your professional experience, education, and other background relevant to the employment opportunity you are seeking. Think of it as highlights on who you are professionally—a summary of your career to date.

The key to writing an effective résumé—one that highlights your best qualifications—is a design that can be scanned in about 20 to 30 seconds. Even within that short amount of time, the prospective employer should be able to glance through your résumé and still have a decent understanding of your background and qualifications.

> **Antimatter power**
>
> High school student Roman Keane has developed a new magnetic exhaust nozzle that would double the velocity of an antimatter-powered rocket. What a résumé builder! It's just that a gram of antimatter, if it could be produced, would cost roughly US$1 trillion.
>
> For details, see the Preface for the URL.

Note   If you are at the beginning of your engineering career, see "Early-Career Résumés" and "Early-Career Application Letters" later in this chapter.

CONTINUOUS REDESIGN AND UPDATE

Developing a résumé is not a one-time effort. Consider it a work in progress: You may have to revise it for every new employment opportunity you seek; you must update it at every accomplishment, milestone, or new phase in your career.

It used to be that a résumé was a fairly permanent record of your background, which you updated only infrequently. You could use roughly the same résumé for many different job searches over a number of years. However, with increasing competition in the job markets and with the availability of desktop publishing software, all that has changed. Now, you may decide to redesign your résumé for every new employment opportunity.

This constant updating is just as important if you settle into one company for a long time. It's easy to forget details about what you've done professionally over the space of just five years. For that reason, keep a working draft of your résumé always at hand—whether as a computer file or as a hardcopy printout on which you scribble notes whenever your career takes a new turn. As of 2008, the joke is that you must keep your updated résumé on file on a CD or flash drive in your coat pocket at all times.

Note   Be aware that most employers now handle résumés primarily by computer—electronically and online. This automated process creates both new opportunities and new hazards. See "Electronic Résumés" in the following pages and Roger Munger, "Technical Communicators Beware: The Next Generation of High-Tech Recruiting Methods." *IEEE Transactions on Professional Communication*, *45*(4) (December 2002).

DESIGN COMPONENTS

The design of a résumé is certainly important to success in an employment search. But a résumé can't do it alone—many other elements have to be present such as connections, timing, need, and of course your actual qualifications. Still, a well-designed résumé does a number of things for you: It highlights your best qualifications, makes it easy for readers to see them quickly, and conveys polished professionalism that reflects positively upon you.

Chronological or Functional Organization.   One of the first issues in résumé design is whether to divide your background information chronologically or functionally. To get a sense of these two organizational approaches to résumés, look at the illustrations in Figure 10-1 for a schematic view and Figure 10-2 for a full-content view.

The *chronological approach* divides your background into education, work experience, and possibly military (although military experience can be distributed into the education and experience sections instead).

One of the strengths of the chronological design is that it shows your work history—in particular, your responsibilities and projects for each organization you've worked for. In the education section, this design shows where you studied and what you studied while there. However, the chronological design does not give a capsule picture of your key qualifications and your key strengths—that information is spread across work and education sections. (One way to solve this problem is to add a highlights section, discussed later in this chapter.)

The *functional approach* divides your background into groups of related education, training, and experience. For example, you may have taken courses in college on project management, attended professional seminars on the subject, taken lead roles in the management of several projects, and maybe even won an award for your management of a project. All of this could be summarized under the heading "Project Management" in a functionally organized résumé (see the schematic illustration of this in Figure 10-1).

The great strength of the functional approach is that it consolidates information about each of your key qualifications, summarizing all relevant work experience and education for each one. From the functional approach, prospective employers looking for someone with project planning and management experience can quickly discern whether you have what they are looking for.

Of course, the weakness of the functional design is the strength of the chronological design: In the functional design, it is immediately clear where, how, and when you

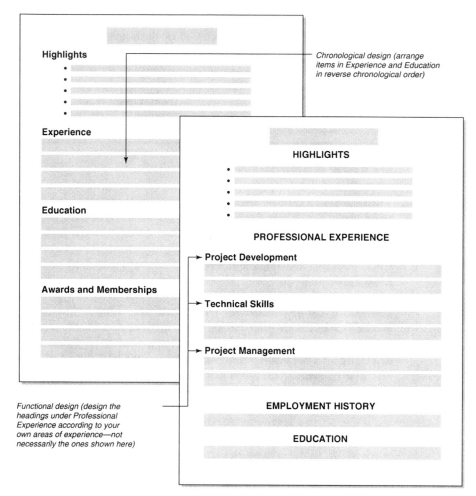

Figure 10-1   Schematic view of example résumé designs. Decide whether the chronological or the functional design works best for you. Visualize the headings you'll use and their relation to each other and to the body text.

gained your experience or education. With the functional design, the chronology of your career must be reconstructed. A solution to this problem is to include a list of your experience and education—no description, just the names and dates (this is schematically illustrated in Figure 10-1 in the heading "EMPLOYMENT HISTORY" and "EDUCATION").

Note   If you are at the beginning of your career, or only a few years into it, consider using the chronological design.

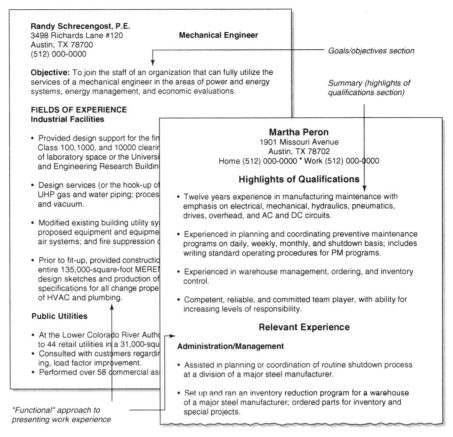

Figure 10-2   Special sections in résumés: the summary or highlights of qualifications sections and the goals and objectives section. A highlights section listing your key qualifications gives a potential employer a quick picture of who you are professionally. Use the objectives section to indicate your professional focus.

Highlights Section.   Another issue is whether to include a highlights section (called different things, including Summary of Experience, Summary, Highlights of Experience, Summary of Qualifications, Synopsis of Qualifications, Professional Expertise, Qualifications, and so on). This section is popular, particularly for professionals who are several years into their careers. It is particularly helpful in résumés that use the chronological design, where key points about your experience and education are scattered throughout the work experience and education sections. Readers have to reconstruct your highlights for themselves.

In the highlights section, however, you do that reconstruction for your readers. The highlights section provides a neat bulleted list of your key accomplishments, key areas of expertise, and key education and training. Even a reader who looks no further in your résumé would still get a picture of who you are professionally.

Notice in Figure 10-2 that a bulleted list is used to make the items in the highlights section more scannable. You position this section just at that point where the eye typically makes first contact with a page. Many believe that our initial glance makes contact with a page one-fourth to one-third of the way down the page, not at the very top. If you believe that, then putting your very "best stuff" at that point in the résumé makes a lot of sense.

Objectives Section. Still another issue in résumé design is whether to include an objectives section. This section describes your career and professional focus. It can indicate the type of work you want to do, the type of position you seek, the type of organization you want to work for, or some combination of these or other objectives.

This section should be brief—no more than two to three lines. It should also be rather specific and not a patchwork of "sweet nothings." For example, avoid this:

Weak objectives statement: Seeking a challenging rewarding career with a dynamic upscale company where I will have ample room for professional and personal growth.

Any reader who was paying attention might ask, "As opposed to what?" Instead, try for something specific:

Improved objectives statement: Construction engineer seeking position in HVAC design and energy calculation for residential and commercial structures.

Some experts argue against the objectives section, fearing that it can narrow your opportunities. However, crafty types rewrite this section to correlate with each position they seek. If it's a large, big-city corporation or a small, rural company specializing in a particular technology, then corresponding words are in the objectives section.

Memberships and Licenses. Another important section in engineering résumés is the list of professional organizations and licenses. In a section like this, indicate that, for example, you are a member of the American Society of Mechanical Engineers.

Specialized Equipment and Knowledge. Many engineers also include in their résumés a section that itemizes their technical knowledge. For example, computer specialists may list the hardware and software they know. Electrical engineers may list their skills in such areas as analog circuit and signal analysis as well as digital and control systems.

Miscellaneous Sections. There are many other possibilities for special sections you can include in a résumé. For example, if you've published articles in professional journals, create a publications section. If you've received honors and awards, create a section for that. If you have received patents, list those in their own section. If you have various security clearances, list them. The idea is to design the résumé so that it emphasizes your best and most important qualifications.

Personal Sections.   Some but certainly not all résumé writers include a section at the very end in which they cite loosely relevant personal details about themselves such as interests, nonwork activities, hobbies, memberships, other languages, and so on. Strictly speaking, this sort of information is out of place in the résumé—what does the fact that you raise orchids have to do with your career as a structural engineer? Viewed more broadly, however, this kind of information rounds you out as a human being. It gives the prospective employer something to chat with you about while waiting for the elevator—to fill those moments of uncomfortable silence.

Presentation of Details.   In addition to planning the overall design and contents of your résumé, you must also decide on how you want to present the actual details of your background and qualifications.

As Figure 10-3 illustrates, there are many ways to show your experience. You can present it in simple paragraphs, as the lowest of the three examples in the figure shows; you can present it in bulleted lists as the other examples in the figure show. You can highlight the name of the organization you worked for by presenting it first in all caps, bold, italics, or bold italics. Or, you can highlight your title or position by presenting it first, as the rightmost example in Figure 10-3 does.

As to the kinds of details you can present in these sections, there are many possibilities, as shown in the following lists. Be selective—don't bury your best qualifications in a mass of less important detail.

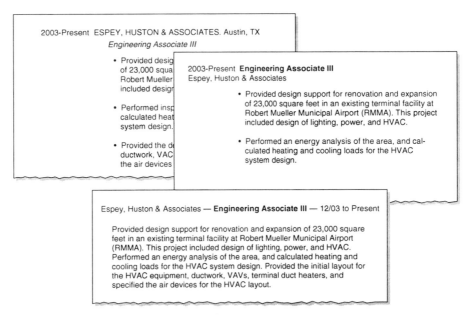

Figure 10-3   Examples of detail formats. Use combinations of list or paragraph format, italics, bold, all caps in the design of the four main elements: date, organization name, job title, and details.

For the experience section, consider including these details:

- Name of the organization where you worked and its address.
- Brief description of the organization, its products, services.
- Your job title and your specific responsibilities.
- Dates of employment with the organization.
- Your major achievements, important projects, promotions, and awards.
- Experience with technologies, equipment, and processes at that organization.

For the education section, here are some ideas:

- Name, location, and brief description of the educational institution.
- Your major and minors, grade point average (overall and in your major).
- Major emphasis of study.
- Important courses taken with descriptions.
- Experience with technologies, equipment, and processes at that institution.
- Important projects.
- Awards, memberships.
- Dates of enrollment and graduation.

> **Batman power**
>
> Engineering students at Brigham Young and Utah State, entrants in the U.S. Air Force Research Laboratory's 2012 Service Academy and University Engineering Challenge, have designed Batman-inspired wall-climbing systems to allow soldiers to safely and quickly ascend vertical surfaces.
>
> For details, see the Preface for the URL.

When you present these details, be as specific as you reasonably can: Cite specific product names, specific measurements and dimensions, specific processes and activities. Consider these examples:

Weak general phrasing	Specific phrasing
Process improvements resulted in considerable savings.	Process improvements resulted in an average cost savings of $315,000 annually.
Work was done to military standards.	Work was done to SAMSO-STD-77-7 military standard.
Redesigned processors for modems.	Redesigned Cy-6000 low-gate processors for QAM/QPSK/FSK-mode modems.

Generalities are less noticeable than specifics; they have far less impact than specifics; and they seem less real, less authentic.

Also, use strong action verbs when you discuss your background and qualifications. Verbs like *designed, developed, utilized, coordinated*, and *supervised* are more striking and memorable than *was involved with, handled*, or *was responsible for*.

OVERALL FORMAT

Figure 10-4 gives you a schematic view of some common ways to design the overall format of résumés. You can see that headings can be centered, they can be placed on the left margin but run into the text, or they can be put in their own column separate from the text. Some résumés add ruled lines horizontally or even vertically to increase the visual separation of the components of a résumé. (See Figure 10-5 for a full-content example of an engineering résumé.)

Format of Headings and Margins.   As you design a résumé, consider the graphic relationship of text to headings. Many résumés use a "hanging-head" design in which the headings are on the far left margin and the body text of the résumé is indented about 1 inch. This design makes the line length of body text shorter and more easily scannable, headings more visible, and the sections of the résumé more visually distinct.

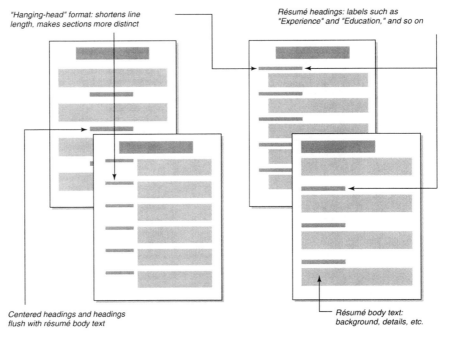

"Hanging-head" format: shortens line length, makes sections more distinct

Résumé headings: labels such as "Experience" and "Education," and so on

Centered headings and headings flush with résumé body text

Résumé body text: background, details, etc.

Figure 10-4   Various possibilities for résumé design. Think about the overall design of your résumé first—how the headings are positioned in relation to the text; visualize it in blocks like these, without the words.

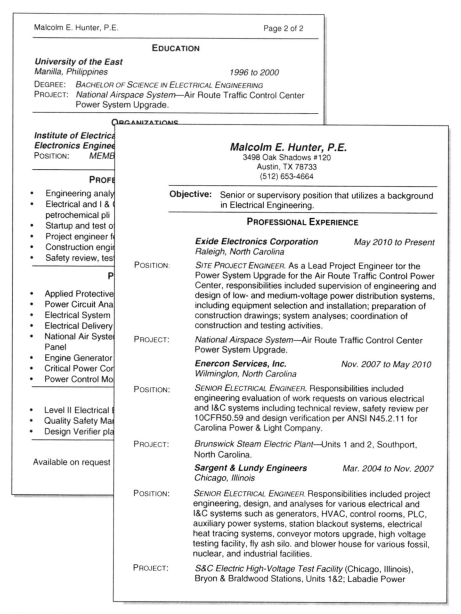

Figure 10-5 Excerpts from the résumé of an experienced professional engineer. Notice the use of small caps for position titles (such as "Site Project Engineer"). The headings on page 2 of this résumé are "Education," "Organizations," "Proficiencies," "Professional Training," "Professional Certification," and "References."

Résumé Length and Headers for Multiple-Page Résumés.   How long your résumé should be depends on how much detail there is in your qualifications. Early in your career you may have trouble filling up a single page. If so, see "Early-Career Résumés" in the following pages. After a few years, however, you'll have trouble keeping your résumé to one page, then two pages, and so on. The chief problem with long résumés is that prospective employers may not read them closely. If you can somehow cut the length of your résumé from three pages to one, the prospective employer is more likely to notice and remember your key qualifications. Some résumé experts maintain that you should plan for one page of résumé for every ten years of experience. However, there are plenty of reasons why this guideline might not be applicable.

In any case, if your résumé is more than one page, place headers at the top of the following pages, as in the examples shown here:

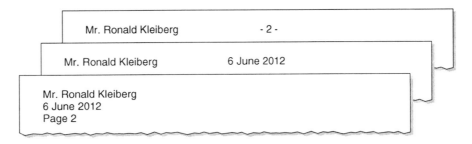

Table 10-1 summarizes these recommendations on writing résumés.

ELECTRONIC RÉSUMÉS

Since the late 1990s employers are increasingly using various electronic and online methods for selecting job candidates. This trend has created much uncertainty, but here are some suggestions.

- **In-house résumé-scanning and -searching applications.** Initially, employers scanned hardcopy résumés and keyword-searched them for job candidates. In this early phase, résumé writers were warned to use a plain, no-frills font and avoid special typographical effects such a bold, italics, alternative font, and different type sizes because of the limitations of scanning equipment. Scanners, however, have improved considerably. Even so, many candidates are expected to send their résumés by email attachment (in electronic form), which eliminates the scanning step altogether.[2]

---

[2]Roger Munger, "Technical Communicators Beware: The Next Generation of High-Tech Recruiting Methods." *IEEE Transactions on Professional Communication*, *45*(4) (December 2002).

Table 10-1   Tips on Writing Résumés

- Include specific details about qualifications and background: specific product names, specific dimensions, specific processes, and specific technologies.
- Use strong action verbs when presenting details about qualifications and background: *designed, developed, coordinated, supervised*, and so on.
- Make sure that the different sections of your résumé are distinct from each other; use spacing, ruled lines, and headings.
- Present education and work experience in reverse chronological order.
- When referring to your own work, omit *I*. Instead of writing "I supervised a team of 12 designers...," write "Supervised a team of 12 designers...."
- Indicate the meanings of abbreviations or acronyms—don't assume the whole world knows what "GPA" or the construction "3.5/4.0" means. Spell out the names of organizations; briefly explain their functions.
- Use format consistently: If you present the details of your work experience using one format, use that same format in other similar areas of your résumés.
- Use consistent margins. Typically résumés have two levels of indentation (for example, in the hanging-head format)—one at which headings align, and another at which text aligns. Make sure all text uses these two levels of indentation and no others.
- Use special typography moderately—for example, bold, italics, underlining, type sizes, and different fonts. Don't go wild with multiple fonts (Times, Helvetica, Thames, etc.) and font styles (bold, italics, underscores, etc.).
- Use special typography consistently: For example, if you put company names in bold in the work-experience section, put college or university names in bold in the education section.
- Keep résumés as short as possible: For example, one page at the start of your career; two pages after you've gained substantial professional work experience.
- Keep the résumé from spilling over by just a few lines to a second or third page. Force the résumé to fill the pages it occupies.
- If your résumé is more than one page, put a header on the second and following pages. Design the header like the ones shown in the preceding pages.
- If you send photocopies of your résumé, get a high-quality photocopy. Request that high-quality paper be used.
- Omit details on age, marital status, sex, religion, handicaps, and other personal matters. Don't include a photograph of yourself.
- Don't omit normal words such as articles (*a, an, the*). Make your writing style compact but not unintelligible.
- Avoid lengthy paragraphs; keep paragraphs under four lines.

- **Job boards and websites.** Another popular option involves "job boards," websites at which job seekers post their résumés for a small fee and at which employers search for candidates for another fee. These can be general such as Monster.com (Monster.com) or America's Job Bank (ajb.com) or specialized such as the National Writers Union Job Hotline (nwu.org) or EngineersforHire.com (www.engineersforhire.com).

Both Internet-based job boards and résumé databases enable employers to search for candidates by using keywords. For example, if the prospective employer were looking for someone with experience in HVAC layout and calculation of heating and cooling loads, these words and their synonyms must appear in the résumé.

- **Corporate websites for recruitment.** As Roger Munger explains in his article on high-tech recruiting methods, employers have not remained satisfied with the methods described in the preceding and have established their own corporate recruitment websites. The practice has become widespread: For example, well over 90% of companies in the manufacturing, healthcare, and high-tech sectors rely on this method. At these websites, employers can fine-tune the online application process to help them identify candidates who actually qualify for the jobs they seek.

- **Blogs and community-oriented Internet applications.** As discussed in Chapter 4 and later in this chapter, blog facilities and community applications enable you to post your résumé on the Internet, thereby making it searchable by potential employers.

- **Online profiles.** Munger points out that employers may gradually abandon the traditional résumé and direct applicants to fill out questionnaires. Applicants' answers will then be used to construct searchable profiles, which provide more detail, a tighter match with employer requirements, and a consistent format.

In the face of all this variability:

- Assume that your résumé will be read, archived, and searched by computers—even for jobs in small companies.

- Carefully follow the application guidelines stated by a potential employer.

- Make sure you send your résumé in the expected format: hardcopy printout or electronic file (PDF); or plain-text ASCII.

- Include a cover letter if requested by the employer, even if you submit your résumé electronically.

- Make sure your résumé contains keywords relating to your qualifications, to the specific job you are seeking, or both. Use industry-standard keywords.

- Use resources—such as journals, newsletters, and conferences—to stay abreast of the evolution of online job searching and online job recruitment.

EARLY-CAREER RÉSUMÉS

If you are at the beginning of your career as an engineer, all the advice and examples to this point may seem fine and good, but what if you have very little experience? Careers must start somewhere—and so must résumés. You can use several strategies to fill out your résumé so that you appear to be the promising entry-level engineer that we all know you are.

- Cite relevant projects (both in academia and community) you've worked on, even if they are not "real" engineering.
- Spend extra time describing college courses and programs you have been involved in. What about team projects, senior projects, or reports?
- Include volunteer work that has had any trace of engineering to it. (If you've not done any volunteer work, get to volunteering!)
- List any organizations you have been a member of and describe any of their activities that have any trace of engineering to them. (If you've not belonged to any engineering-related organizations, get to belonging!)
- Use formatting to spread what information you have to fill out the résumé page.

In the student résumé shown in Figure 10-6, notice how much space that details about education take up. This résumé writer could have included even more: Descriptions of key courses and projects could have been provided under a heading such as "Essential Coursework."

Notice too that the résumé in Figure 10-6 includes plenty of co-op and part-time work. The bulleted-list format extends the length of the résumé so that it fills up the page. At the bottom of the résumé, the writer lists awards and organizations. These too could be amplified if necessary. Details as to what the award is about, why this writer received it, and what those organizations are—these are examples of good information that could be added, if necessary.

Subtle changes in format can help make your résumé fill a page. Top, bottom, left, and right margins can all be pushed down, up, and in from the standard 1.0 inch to 1.25 inches. You can add extra space between sections. To do so, don't just press Enter or Return. Instead, use the paragraph-formatting feature of your software to put 6 or 9 points, for example, below the final element of each section. Line spacing is another subtle way to extend a résumé. If your software by default uses 13.6 points of line spacing for Times New Roman 12 point text, experiment with changing the line spacing to exactly 15.0 points.

> **Greenhouse on Mars**
>
> In 2002, students at Olin School of Engineering and elsewhere entered their designs for a deployable Mars greenhouse in the NASA's Mars-Port Engineering Student Design Competition.
>
> For details, see the Preface for the URL.

HOW TO WRITE AN APPLICATION LETTER

Often accompanying the résumé is the application letter, sometimes called a cover letter. This letter is the first thing that potential employers see when they open the envelope—the application letter on top, with the attached résumé beneath it.

Whether to include an application letter with your résumé depends on the potential employer. Sometimes, only the résumé is requested; sometimes, only the letter.

ROBERT McILWAIN
2009 Thistle Bluff
Austin, TX 78713
(000)000-0000
egibbon@rome.ece.utexas.edu

Objective	Employment as an entry-level VLSI design engineer in research and development.
Education	The University of Texas–Austin GPA: 3.9/4.0
	Electrical and Computer Engineering
	Bachelor of Science degree expected August 2013
	Related Courses: Digital system design using VHDL, operating system development using MC68340, computer-aided VLSI design using Magic, and DLX RISC microprocessor design using Compass.
Skills	Proficient in assembly language (MC6800 family), C++, and VHDL. Proficient in UNIX and system administration. Skilled in simulation tools including Workview, PSPICE, and Compass.

Experience

09/04 – 12/04   *Engineering Co-op, National Systems*   Austin, Texas
- Analyzed and documented customer problems.
- Designed and verified hardware parts.
- Tested Windows 7 device driver.

12/03 – 08/04   *Assistant System Administrator, University of Texas–Austin*
- Administrated a network with over 40 users.
- Upgraded operating system from SunOS to Solaris.
- Designed homepage for the Telecommunication Laboratory.

12/02 – 05/02   *Engineering Co-op, National Systems*   Austin, Texas
- Developed and implemented verification plan for new General Purpose Interface Bus hardware product and software application.

06/01 – 08/01   *Engineering Co-op, National Systems*   Austin, Texas
- Developed hardware diagnostic program.
- Designed test setup using SBus expansion chassis.

01/01 – 5/01   *Grader, The University of Texas–Austin*
- Assisted professor in grading student homework in entry-level circuit theory class.

Honors & Activities	Recipient of Engineering Scholar Award Member of Tau Beta Pi Engineering Society Member of Eta Kappa Nu Electrical Engineering Honor Society
References	Available on request

Figure 10-6   Résumé of a graduating engineering student.

Sometimes, after prospective employers make their initial selection of candidates, they request the other of the two components.

There are two categories of application letters, based on the information they contain:

- **Cover letters.** In the true cover letter, you simply announce that a résumé is attached, indicate that you are investigating an employment opportunity,

6307 Marshall Lane
Austin, TX 78703

25 May 2012

Ms. Juanita Jones
Hughes & Gano, Inc.
P.O.Box 1113
Austin, TX 00000

Dear Ms. Jones:

Please accept the attached résumé as my application for the
position of Process Engineer currently available with your
company.

I'll be looking forward to meeting with you at your earliest
convenience. I can be reached at (512) 471-4991 during regular
working hours or at (512) 471-8691 in the evenings and
weekends.

Please contact me if you need any further information about my
background or qualifications.

Sincerely,

Patrick H. McMurrey
Encl.: résumé

Figure 10-7   Cover letter: A brief correspondence that identifies
the position being sought and the purpose of the correspondence.
For most job searches, use the full application letter, as described in
this chapter.

and specify the position you seek. As illustrated in Figure 10-7, this is a
brief letter, the body text totaling less than 10 lines. If the job advertisement
asks for résumés only, you can still include this type of letter to identify the
position.

- **Full application letters.** In the true application letter, you discuss your background and qualifications as relevant to the position you are seeking. The job
  of this letter is to promote yourself—to highlight the reasons why you are right
  for the position. This type of letter is the focus of the rest of this chapter.

Which to use? The cover letter is certainly easier to write, but it doesn't do anything
for you. The full application letter acts as your proxy, showing the prospective employer
specifically which of your qualifications make you right for the job. If a full application
letter is expected, sending only a cover letter makes you seem noncommittal or even
indifferent.

Note   If you are at the beginning of your engineering career, see "Early-Career Application Letters" later in this chapter.

Application Letter: Contents and Organization

The function of the application letter is to introduce you to the prospective employer, state the purpose of the letter (to seek an interview), identify the position you're inquiring about, and summarize your relevant qualifications. This last function is the most important. The application letter is not just another form of the résumé—it is a careful selection from the résumé. It makes a strong case for you as a good candidate for the specific position by pointing out aspects of your background that are a good fit with the specific job you are seeking.

First Paragraph.   The first paragraph of the letter should be brief and do some combination of several things: state the purpose of the letter (to inquire about employment); state how you found out about the opening, if applicable; say something that will catch readers' attention and make them want to continue reading—and that's it. Keep this first paragraph short, four lines at the maximum.

In this first paragraph, consider using one of several common tricks to catch readers' attention:

- State something specific about your qualifications that makes you the right candidate for the position.
- Cite information about the company to which you are applying—information that shows you are informed and that relates to the position you seek.
- If possible, mention the name of someone in the company who knows you and can speak favorably about you.
- Say something enthusiastic or energetic about the kind of work you want to do, the kind of organization you want to work for, or maybe something about your professional goals.

Whichever of these strategies you use, remember to keep it short. Also, remember to write in terms of the potential employers' perspective. For example, employers don't want to hear at length about how much you look forward to sailing on the nearby lake.

Middle Paragraphs.   The middle portion of the letter discusses your qualifications that relate specifically to the employment you seek. Somewhere in these paragraphs, suggest that readers see the attached résumé for more detail. In these paragraphs, use the same kinds of organization as in the résumé.

- **Chronological approach.** Discuss your education in one or more paragraphs, and then your work experience in another set of paragraphs. If work experience is your best "stuff," put it before education.

- **Functional approach.** Focus on the important areas in your qualifications—for example, project management, research and development, quality control, vendor coordination, or documentation. Ideally, reserve a separate paragraph for each of these areas. In each, discuss anything in your background—whether work experience, training, awards, or education—that relates to that area.

These organizational approaches are schematically illustrated in Figure 10-8. If the middle paragraphs take up too many lines, consider using a bulleted list (as illustrated in Figure 10-9). This format enables you to present important details but in a condensed, more readable and scannable way.

Final Paragraphs. In the final portion of the letter, you wrap it up: Mention that the résumé is enclosed if you've not already done so; urge the prospective employer to get in touch; facilitate arrangements for an interview; and find some parting encouraging, enthusiastic comment to make, such as your strong interest in the employment, the company, the profession, and so on. Some job seekers indicate that they will call the prospective employer on a certain date (for example, a week after mailing the letter). While others might find this tactic too aggressive, it certainly puts pressure on the employer to take action.

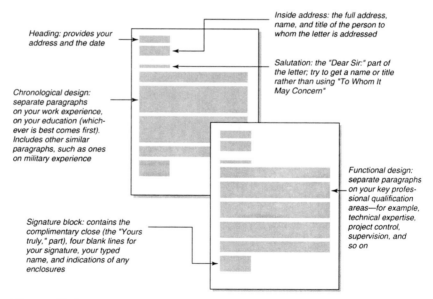

Inside address: the full address, name, and title of the person to whom the letter is addressed

Heading: provides your address and the date

Salutation: the "Dear Sir:" part of the letter; try to get a name or title rather than using "To Whom It May Concern"

Chronological design: separate paragraphs on your work experience, on your education (whichever is best comes first). Includes other similar paragraphs, such as ones on military experience

Functional design: separate paragraphs on your key professional qualification areas—for example, technical expertise, project control, supervision, and so on

Signature block: contains the complimentary close (the "Yours truly," part), four blank lines for your signature, your typed name, and indications of any enclosures

Figure 10-8 Common sections of application letters. You can organize the letter chronologically or functionally the same as you can the résumé.

Patrick H. McMurrey

1108 West 29

Austin, TX 78703

(000) 471-0000 (home) — (000) 000-0000 (work)

May 25, 2012

Director of Personnel
Automation Associates
7805 Pearl Creek Drive
Austin, Texas 78706

Dear Director of Personnel:

I would appreciate your time in evaluating my qualifications in relation to your current needs for a Senior Electrical Engineer in Automation Associates's large building design projects. Attached is a copy of my current résumé.

I have over 15 years experience in various facets of electrical design and engineering. Specifically, I have experience in power and control design including analyses for power generation; low-, medium-, and high-voltage power distribution systems; fire detection and protection systems; plant security systems, programmable logic controllers (PLCs); as well as equipment layout for various types of industrial facilities.

I am currently employed with Exide Electronics Corporation as Site Project Engineer in the National Air-Space Federal Systems Engineering Division. My current responsibilities are as follows:

- Lead project engineer for the power system upgrade of Denver, Albuquerque, Indianapolis, and Jacksonville Air Route Traffic Control Centers (ARTCC). This power upgrade is part of Exide's current contract with the U.S. Air Force and the Federal Aviation Administration.

- Supervision of varying numbers of electrical engineers and designers in various engineering tasks for projects such as low- and medium-voltage power distribution system engineering and design.

- Vendor interface for installation of equipment such as diesel generators, switchgears, and power control monitoring systems.

This current work and past projects, along with the references I am including in this letter, all attest to my solid record of initiative, responsibility, creativity, and professional dedication. I am an effective, contributing member of any organization that I am associated with. If you are interested in discussing my experience and capabilities further, please contact me at one of the numbers shown above.

Sincerely,

Patrick H. McMurrey
Encl.: résumé, reference list

Figure 10-9 Example of an application letter: Notice how much specific detail the writer packs in concerning his experience. Notice also how the bulleted list relieves some of the density of the letter.

FORMAT OF APPLICATION LETTERS

As for the format of the application letter:

- Use a standard business-letter format, such as the one shown in the examples in this chapter. (See Chapter 4 for style and format of business letters in general.)
- Single-space the individual components (never double-space). Double-space between the components.

Leave four blank lines between the complimentary close and your typed name, and sign your name in that space.

- Do not indent the first line of paragraphs of the body text. Use standard left and right margins; 1 inch, 1.5 inches, or 2 inches are all acceptable. Use wider margins when your letter seems too skimpy.
- Use standard top and bottom margins: The letter can begin anywhere from 1 inch to 3 inches from the top edge of the page; it should end no closer than 1 inch to the bottom edge.
- Carefully position your letter on the page. If your letter is short, use the variables of margins and spacing between text components to position the text of the letter in the upper middle of the page.
- Avoid dense paragraphs. Don't expect readers to labor through paragraphs over eight or more lines. Use paragraph breaks and numbered or bulleted lists to loosen up dense paragraphs.
- For additional eye appeal, consider creating an attractive, professional-looking design for your name and address, such as those illustrated in Figure 10-10.

Tone in Application Letters.   In an application letter, tone can also be an important characteristic, but also hard to define. Tone should reflect your view of yourself and the type of professional you want to be. You may want to avoid sounding brash, arrogant, or overconfident—unless that's your personality. The following examples explore how bad tone can result from good intentions:

- **Stiff, overly formal, overly reserved.** When you write an application letter, the pressure is on—obviously. One tendency is to freeze up and be overly cautious. Ironically, this can sound like indifference or create a stiff, reticent, overly formal tone—a personality prospective employers would rather avoid.
- **Intimidatingly qualified, or even overqualified.** Tone can also go bad when you overemphasize your qualifications and make yourself sound like a miracle worker. Employers may get uneasy—fearing the prospect of a co-worker with an overdeveloped ego. They may worry about the safety of their own jobs, or they may wonder whether you're stretching the truth or simply lying.
- **Unctuous, fawning, flattering.** It's possible to try too hard to sound bright, positive, enthusiastic, and eager; it's possible to sound phony in saying nice things about the prospective employer.

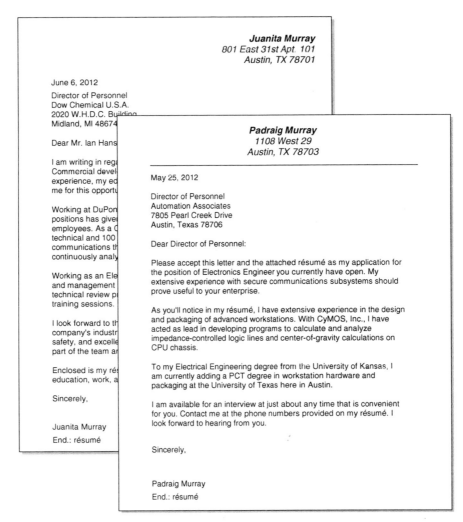

Juanita Murray
801 East 31st Apt. 101
Austin, TX 78701

June 6, 2012

Director of Personnel
Dow Chemical U.S.A.
2020 W.H.D.C. Building
Midland, MI 48674

Dear Mr. Ian Hans

I am writing in rega
Commercial devel
experience, my ed
me for this opportu

Working at DuPon
positions has give
employees. As a C
technical and 100
communications th
continuously analy

Working as an Ele
and management
technical review p
training sessions.

I look forward to th
company's industr
safety, and excelle
part of the team ar

Enclosed is my ré
education, work, a

Sincerely,

Juanita Murray
Encl.: résumé

Padraig Murray
1108 West 29
Austin, TX 78703

May 25, 2012

Director of Personnel
Automation Associates
7805 Pearl Creek Drive
Austin, Texas 78706

Dear Director of Personnel:

Please accept this letter and the attached résumé as my application for
the position of Electronics Engineer you currently have open. My
extensive experience with secure communications subsystems should
prove useful to your enterprise.

As you'll notice in my résumé, I have extensive experience in the design
and packaging of advanced workstations. With CyMOS, Inc., I have
acted as lead in developing programs to calculate and analyze
impedance-controlled logic lines and center-of-gravity calculations on
CPU chassis.

To my Electrical Engineering degree from the University of Kansas, I
am currently adding a PCT degree in workstation hardware and
packaging at the University of Texas here in Austin.

I am available for an interview at just about any time that is convenient
for you. Contact me at the phone numbers provided on my résumé. I
look forward to hearing from you.

Sincerely,

Padraig Murray
Encl.: résumé

Figure 10-10 Examples of application letters: The first paragraph of the letter on
the right identifies the position being sought and makes one strong statement about
the writer's qualifications. (The fancy headers are not a requirement—just a nice,
professional-looking, eye-catching touch.)

- **Hungry, desperate.** Avoid the tone that says "I'll do anything!" The anxiously
  eager tone can go bad when it starts sounding desperate for a job—any job.
  Maintain your professional focus and integrity—you won't do just any kind of
  work; you want the kind of work you have trained for.
- **Overly humble, overly simple, above it all.** It might be tempting to adopt the
  attitude that says "this is who I am, this is what I can do, this is what I have
  done—take it or leave it." It's simple, humble, plain, no-nonsense. But it can

sound so excessively (even aggressively) humble that employers may decide the job seeker will prove unbearably superior, even arrogant.

Bad tone can start from good intentions: You certainly want to be cautious and respectful; to show what's good about yourself; to be enthusiastic and complimentary; to sound comfortable and confident professionally; to demonstrate that you are earnest about the employment opportunity; and to be honest and straightforward. But if you handle any of these strategies clumsily, problems of tone occur and you run the risk of projecting the wrong image of yourself.

Table 10-2 reviews the guidelines in this chapter on application letters.

> ### *Speed record for student-built electric car*
>
> In 2011, Electric Blue, built by Brigham Young manufacturing engineering students, set a record of 205 km/h in the E1 class of electric cars (under 499 kg). The previous record was 201 km/h.
>
> For details, see the Preface for the URL.

EARLY-CAREER APPLICATION LETTERS

In the preceding, you've seen some rather impressive application letters. But what if you don't have all that experience—how do you construct a respectable application letter? It's the same problem addressed earlier in "Early-Career Résumés," and most of the strategies are the same.

- Describe relevant projects (both in academia and community) you've worked on, even if they are not "real" engineering.
- Spend some time describing essential college courses and programs you have been involved in. What about team projects, senior projects, or reports?
- Include volunteer work that has had any trace of engineering to it. (If you've not done volunteer work, get to volunteering!)
- Describe any organizations you have been a member of that have any trace of engineering to them. (If you've not belonged to any engineering-related organizations, get to belonging!) Describe the engineering-oriented activities of those organizations.
- As with the résumé, you can use formatting to spread what information you have to fill out the résumé page. See "Early-Career Résumés" for strategies.

In the example student application letter in Figure 10-11, notice that the writer describes his coursework and the applications that he used. His reference to a professional exposition shows an active interest in a particular area of the engineering profession. Moreover, his visit with an employee of the company with which he seeks employment is a crafty form of name dropping. In general, the letter expresses enthusiasm about working in the VLSI area.

Table 10-2   Tips on Writing Application Letters

- Avoid diving headlong into the details of your background and qualifications in the very first paragraph. Create an introductory paragraph that performs the functions mentioned earlier in this chapter.
- Get a specific name or department to which to address the letter; avoid the "To Whom It May Concern" syndrome.
- Individualize the letter for the addressee. Even if you are in a massive job search and are sending out many letters, avoid sounding as though you're a zombie writing form letters.
- Be sure to mention that your résumé is enclosed with the letter.
- Use standard business letter format in the application letter, as shown in the examples in this chapter and as described in detail in Chapter 4. (Remember to punctuate the salutation with a colon, not a comma!)
- Keep the letter to one page. Keep the paragraphs of the letter short: the first paragraph under five lines; the body paragraphs under eight lines.
- Seek a nice, bright, energetic, positive tone. Watch out for the problems with tone discussed in this chapter. (Get someone to read a rough draft of your letter and describe the kind of personality it projects.)
- Write the letter in terms of the prospective employer's needs or interests, and only minimally your own. Discuss yourself according to the prospective employer's needs.
- Use the full application letter (as opposed to the cover letter) unless the job advertisement specifically requests only the résumé.
- Avoid negative discussion of previous employers; generally avoid stating reasons why you left previous jobs.
- Unless specifically requested by the prospective employer, avoid discussion of salary, benefits, or other compensation.
- While it's acceptable to send out high-quality photocopies of the résumé, the letter should be freshly printed out. Make the letter appear as though you prepared it especially for the addressee.
- Avoid spelling, grammar, usage errors, and bad writing at all costs!

USING THE INTERNET FOR JOB SEARCHES

You are probably aware that many professionals now use the Internet to seek employment and display their professional qualifications. Things started with individuals creating their own websites that included their résumés. Soon, résumé- and job-posting services, such as Monster.com, sprang up. Some of these are free; others you pay to post your résumé.

More recently, however, people are using blogs and "social-networking" facilities (such as LinkedIn) to put their qualifications out on the Internet. The advantage of

Edward Damien
1307 Marshall Lane
Pflugerville, TX 78660

June 6, 2013

Vern Whittington
University Recruiting Manager
Dallas Semiconductor
4401 South Beltwood Pkwy
Dallas, TX 75244-3292

Dear Mr. Whittington:

I am writing you to express my interest in becoming a VLSI design engineer with Dallas Semiconductor. I will earn my BS degree in Electrical Engineering from The University of Texas at Austin in August, 2013. My objective upon graduation is to become a successful VLSI design engineer in the semiconductor industry.

During the Engineering Career Exposition in September 2012 Tiffany Oberlin, a Dallas Semiconductor college staffing coordinator, talked to me about career opportunities with Dallas Semiconductor. Her description of the company's wide range of products, especially touch memory for automatic identification, impressed me. I am very interested in becoming part of the VLSI design team working on this challenging project.

As my enclosed résumé explains, I have completed courses related to VLSI design, including digital system design and reduced instruction set microprocessor design. I am also proficient in several VLSI design tools such as Synopsys and Workview. In addition, my three co-op tours with National Instruments have demonstrated my ability to work with people and to apply my technical knowledge to practical tasks.

I am looking forward to discussing my qualifications with you. Please feel free to contact me either at (512)111-2222 or at platapus@aussieu.edu. Meanwhile, I greatly appreciate your kind help and attention.

Sincerely,

Edward Damien

Enclosure

Figure 10-11   Example of an application letter of a graduating engineering student.

using these resources is that you don't have to create your website or know anything about XHTML.

To explore how engineers are using blogs and social networking, try these two ideas:

- To see engineering blogs, go to Google.com, click More... and then search on engineers. As of this publication, the search results yielded mostly job announcements, but blogs of individual engineers (displaying their professional qualifications) were scattered among these (see Figure 10-12).

- To see online engineering profiles, go to LinkedIn.com, and search on engineers. You'll see an extraordinary number of engineers who have built a professional presence in facilities like LinkedIn (see Figure 10-13).

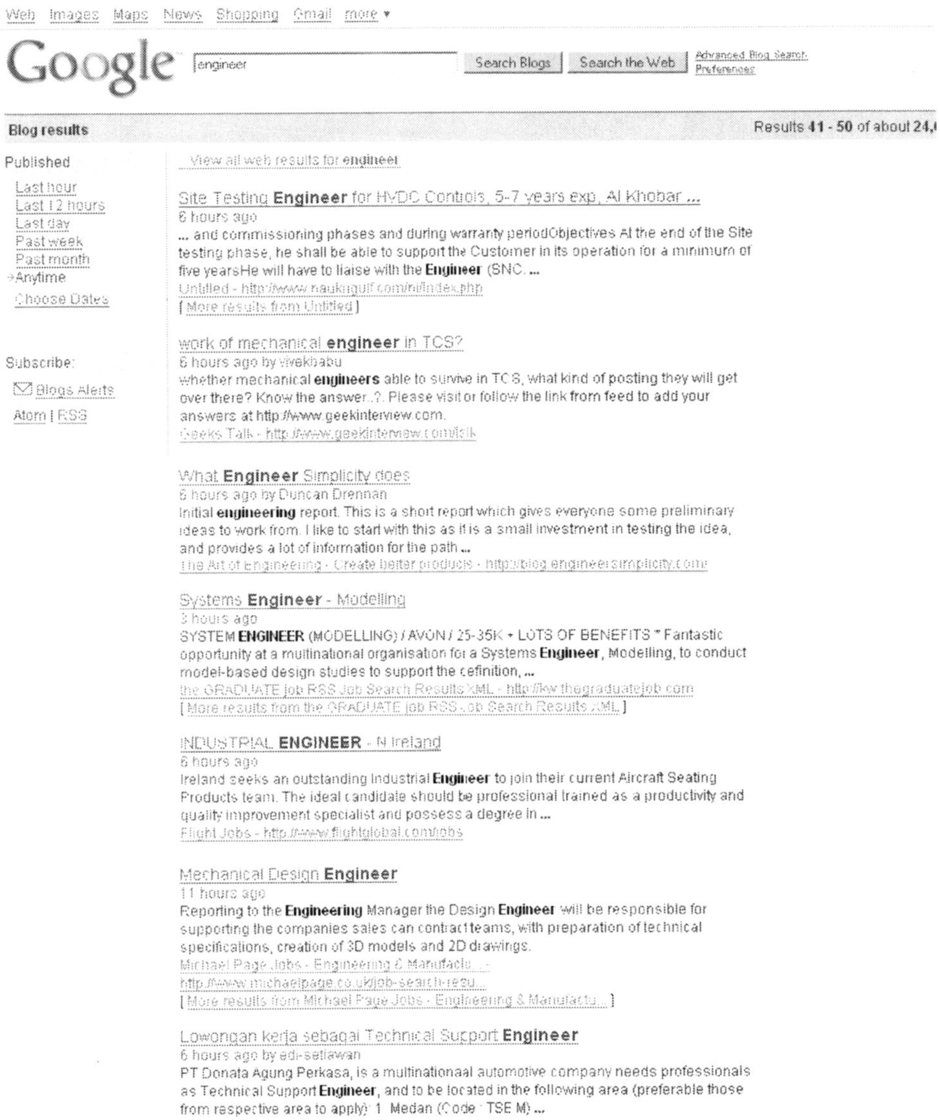

Figure 10-12   Results of a search for "engineer" at Google.com. While most of the results are job announcements, plenty of engineers maintain their own blogs to display their qualifications or discuss professional issues.

Figure 10-13   LinkedIn profile page of a professional engineer. Adapted with permission.

HOW TO WRITE A FOLLOW-UP LETTER

Write a follow-up letter when you've not heard from a prospective employer after two weeks, after you've had an interview, when you want to acknowledge a refusal of a job offer, and when you must reject or accept a job offer. The most important use of the follow-up letter is for those situations when you are waiting (and waiting) and have had no word from the prospective employer. (See Figure 10-14 for an example.) To write a follow-up letter, consider including these contents:

- State the reason you are writing the letter—to inquire about the application letter and résumé you recently sent.
- Indicate the date you sent the letter and the résumé and specify the position you were inquiring about.

801 East 31st Street #101
Austin, Texas 78701

3 March 2012

Director of Personnel
Automation Associates
7805 Pearl Creek Drive
Austin, Texas 78706

Dear Director of Personnel:

On February 17, I applied for a position as manufacturing
engineer with your firm. Not having heard from you in the two weeks
since that time, I'm concerned that my letter may have been lost.

Attached is a copy of the original letter and résumé that I sent. As
you will see, they detail my work experience, my education, and my
sincere interest in working for your company.

If you have already made a decision, I would appreciate hearing from
you. For the moment, my availability continues. I look forward to
discussing the job and my background with you in person.

Sincerely,

Juanita Murray

Encl.: Copy of 2-17 letter and résumé

Figure 10-14   Follow-up to an application letter: Its most
important use is to inquire about the fate of an application letter
and résumé for which you have received no response.

- Suggest that the letter and the résumé might have been lost in the mail (or email) or routed incorrectly within the recipient's organization.
- Enclose a copy of the original letter and résumé and state in the follow-up letter that you have enclosed them.
- Tactfully encourage the recipient to let you know the status of the position (indicating that your own decisions are dependent upon it).

Students to save Manhattan

University of Pennsylvania students have designed waterproof canopies to protect against rising sea levels, which could rise by up to 21 cm depending on future greenhouse gas emission levels.

For details, see the Preface for the URL.

EXERCISES

Talk with several professional engineers about the application letters and résumés they typically see when hiring new engineers:

1. How do they "read" résumés: line by line from beginning to end? If they skip around and scan, what do they look for? What catches their eye? How important are specific details such as brand names, model numbers, titles of specifications, and dimensions?

2. What can the engineer who is just graduating and getting started in the profession legitimately put in the work experience section of a résumé?

3. Should personal information such as hobbies, community activities, or reading interests be kept strictly out of résumés? If not, what purpose do they serve?

4. What are the typical problems that cause a résumé to be ignored? How much does the formatting of a résumé contribute to their willingness to read a résumé carefully?

5. Are applicants asked to send only a résumé or only an application letter? Do they expect to see a simple cover letter (as described in this chapter), or do they expect a full application letter?

6. Does tone ever cause a problem in these letters?

BIBLIOGRAPHY

Munger, Roger. Technical Communicators Beware: The Next Generation of High-Tech Recruiting Methods. *IEEE Transactions on Professional Communication*, 45(4), December 2002.

U.S. Department of Labor, Bureau of Labor Statistics. Search for your area of engineering at www.bls.gov/ooh/About/Career-Guide-to-Industries.htm

11

ETHICS AND DOCUMENTATION IN ENGINEERING WRITING

Technology has a pervasive and profound effect on the contemporary world, and engineers play a central role in all aspects of technological developments. In order to hold paramount the safety, health, and welfare of the public, engineers must be morally committed and equipped to grapple with ethical dilemmas they confront.

Mike W. Martin and Roland Schinzinger, *Ethics in Engineering*, 4th ed. (New York: McGraw-Hill, 2004), p. *xv*.

If it is not right do not do it; if it is not true do not say it.

Marcus Aurelius, 121–180 CE.

ENGINEERING ETHICS

It's hard to live very long without making numerous decisions affecting both our own well-being and that of others. Engineers are no different from anyone else: During your career you will have to make countless choices among various courses of action. Thus, you should be familiar with some of the factors involved in making ethical choices. You should also have an idea of the kinds of situations that will require you to make responsible decisions as an engineering writer. After reading this section, you will be aware of pitfalls to avoid as a writer and resources to help you avoid them.

Note   The material on engineering ethics contained in this section is only a general introduction. Much more could be said on the topic, and many books and websites go far more extensively into what we touch on here. Workshops, courses, and distant learning are also available from many sources. An excellent starting place for those who wish to go deeper into this vital part of an engineering career is Texas Tech University's Murdough Center for Engineering Professionalism at www.niee.org

Wherever you find technology, you find ethical and moral concerns. For example, manufacturing and selling an automobile when it is known to be unsafe is an ethical, as well as a legal matter. Where to dump hazardous waste raises considerable moral questions, as do the issues, for example, of building with asbestos or locating high-power transmission lines. Accurate record keeping, the ethical use of software, or professional consulting outside of your regular job are also examples of situations where you as an engineer might find yourself making ethical and moral choices. While working with other people, you may also at times be confronted with issues of dishonesty, discrimination, harassment, and alcohol or drug abuse—all situations calling for sound ethical decisions.

> **Sonar technology deafening cetaceans?**
>
> The UK's Whale and Dolphin Conservation Society alleges that low-frequency military sonar technology used in the search for oil and gas is causing hearing loss in cetaceans (whales, dolphins, and porpoises), injuring them and causing them to strand themselves. The February 2011 mass stranding of over one hundred whales on the coast of New Zealand is seen evidence.
>
> For details, see the Preface for the URL.

FIVE COMMUNICATION CONCERNS

This section focuses on five concerns you must be aware of as an engineering writer and researcher. Some of these concerns are actually illegal practices engineers sometimes commit either knowingly or unknowingly. In some instances they have paid heavy prices for their actions, such as lawsuits, job loss, or at least a diminished reputation. Here are five major areas where problems can arise for engineers who produce information to be shared with others in writing.

Copyright Infringement.   Just because an image or article is available in print or on the Internet does not mean that anyone has a right to copy and use it. If you come up with original ideas or inventions as a result of your own research, copyright them so that they are protected under law. A copyright is the legal right you (or your company) is granted to enjoy complete possession and profits from your work for a certain time. To obtain a copyright, file a copyright office form, pay a filing fee, and provide the copyright office with one or more copies of the work to be copyrighted. Once you have copyrighted your work, it cannot be used or distributed without your permission (with a few minor exceptions). If someone does so, they have infringed on your copyright and you may be able to sue them.

While the previous may be an oversimplification, you can find complete information at the U.S. Copyright Office website at www.copyright.gov. For the engineering researcher and writer, the important point is *always* to be aware of what is someone else's intellectual property and to never use it or cite it in any way without permission or acknowledgment. The exception to this is that you don't need permission to quote or paraphrase a small amount of copyrighted work for educational purposes, as long as you give credit to the source and gain no financial profit from your use of the work. Also, you don't need permission to cite or borrow material from U.S. government publications.

Tampering with Results. Engineers often have to write up the results of their research and experimentation. What if the numbers don't quite come out the way they were supposed to? For example, your team is working on a suspension bridge and has run into a small problem toward the end of construction. The team decides the problem can be overlooked if a few measurements are changed to meet requirements. In your final report, would you carefully change a few numbers so that things "come out right"?

Confronted with such issues, an ethical engineer wouldn't change any results and would work until the problems had been solved and everything was accurate. Sometimes it might seem a few changed details won't hurt, but tampering with results is a very serious issue in the engineering field and is a choice that sooner or later can come back to haunt you.

Another form of tampering with results is found in concocting data. Here a writer makes up information or results with no backing or truth behind them—they are fictitious. Unethical engineers (and other professionals) have been known to insert concocted data in reports to show progress or results that are nonexistent, often in order to get further funding or to hide a lack of real effort. Again, time and suspicion have a way of uncovering such actions.

Withholding Adverse Information. Plenty of engineering evidence shows that withholding adverse information can lead to problems, accidents, and even deaths. Ford Pintos and Firestone tires immediately come to mind in this context. If any kind of damage results because you withhold information about a flawed design, a dangerous product, or a means to avoid harm, it is your or your company's responsibility. You can certainly be held liable for your inaction. No ethical engineer should keep silent or fail to include in a written report anything concerning a product or process that might result in a user's financial loss, physical harm, or death.

Withholding adverse information can also occur in job applications and résumés. It is a temptation to omit less admirable events in one's past, just as it is a temptation to concoct data for them. As several well-publicized cases in recent years have shown, many companies and institutions maintain strict policies that enable them to fire workers who falsify résumés in any way. Instead, focus on your strong points. Thus, you are helping yourself when you write these—and all—documents in an ethical manner.

Writing Unclear Instructions.   In your engineering career, you may get involved in writing instructions, procedures, manuals, or user guides at some point. These must be written in a detailed and precise manner, without error or ambiguity. Imagine the results of unclear instructions for operating an aircraft, space shuttle, or nuclear reactor. Imagine the results of ambiguity about assembling or operating everyday products—such as computers, cameras, pumps, filters, or telescopes: frustration and anger, plus a diminished respect for the product and the company that produced it.

Examples of unclear instructions and their consequences abound. The problem is typically poor planning on the writer's part, careless or hasty writing and editing, or a failure to put oneself firmly in the head of the reader or user. See the section on instructions in Chapter 5 for some good background on writing effective instructions. The more skilled you become at producing watertight instructions, the less likely you will be to frustrate your readers, anger them, endanger them, or be sued by them.

Omitting Safety Warnings.   Engineers should constantly be concerned with the safety of their customers and of anyone else their products and designs might affect. This means you must include clear and conspicuous safety warnings in any design, procedure, or product that requires them. You are always responsible for providing information that ensures the consumer's safety. Failure to provide adequate safety warnings can lead to mishaps, loss, disaster, serious physical harm, or even death. Always take great care to provide clear safety warnings whenever necessary in your writing—and ensure they are visually prominent and accessible to your reader. See the section on instructions in Chapter 5 for one common hierarchy of safety warnings.

> ### *Fracking causing pollution?*
>
> Organizations like foodandwater-watch.org believe that fracking (the injection of massive volumes of water, sand, and chemicals underground at high pressure to break up rock formations thus allowing oil or gas to flow up the well) pollutes air and water. The petroleum industry vigorously disagrees.
>
> For details, see the Preface for the URL.

TOOLS FOR ETHICAL DECISION MAKING

Faced with any of the above problems, you can use some tools to justify doing the right thing. Some of the most powerful are the Codes of Ethics published by professional engineering associations and by some of the larger engineering firms. You can find many of them online by entering "code of ethics of engineers." One excellent such source is The Online Ethics Center for Engineering and Science at Case Western Reserve University, which maintains an extensive listing of codes of ethics for engineers and scientists from around the world. With such documents in hand, you can refer to guidelines that will support your decisions to hold out for strictly ethical writing (and other activities) as an engineer.

Accreditation Board for Engineering and Technology

CODE OF ETHICS OF ENGINEERS

THE FUNDAMENTAL PRINCIPLES

Engineers uphold and advance the integrity, honor, and dignity of the engineering profession by

I. using their knowledge and skill for the enhancement of human welfare;

II. being honest and impartial, and serving with fidelity the public, their employers, and clients;

III. striving to increase the competence and prestige of the engineering profession; and

IV. supporting the professional and technical societies of their disciplines.

THE FUNDAMENTAL CANONS

1. Engineers shall hold paramount the safety, health, and welfare of the public in the performance of their professional duties.

2. Engineers shall perform services only in the areas of their competence.

3. Engineers shall issue public statements only in an objective and truthful manner.

4. Engineers shall act in professional matters for each employer or client as faithful agents or trustees, and shall avoid conflicts of interest.

5. Engineers shall build their professional reputation on the merit of their services and shall not compete unfairly with others.

6. Engineers shall act in such a manner as to uphold and enhance the honor, integrity, and dignity of the profession.

7. Engineers shall continue their professional development throughout their careers and shall provide opportunities for the professional development of those engineers under their supervision.

ABET

345 East 47th St., New York, NY 10017

1987

Figure 11-1   A typical code of ethics for the engineering profession. Use documents like this to support your position when faced with an ethical choice of action.

Two such codes are shown in Figures 11-1 and 11-2. Use them as support if you must defend your decisions. Following the codes of ethics is a suggested checklist for ethical decision making (Figure 11-3) that you might also find useful when you are uncertain about choices or plans of action.

PERSONAL ETHICS AND YOUR CAREER

Our personal ethics are often determined by our personal philosophy of human existence. Do you feel the prime goal of human endeavor is simply to survive or to

---

THE INSTITUTE OF ELECTRICAL AND ELECTRONICS ENGINEERS, INC.

Code of Ethics

We, the members of the IEEE, in recognition of the importance of our technologies in affecting the quality of life throughout the world, and in accepting a personal obligation to our profession, its members and the communities we serve, do hereby commit ourselves to the highest ethical and professional conduct and agree:

1. to accept responsibility in making engineering decisions consistent with the safety, health, and welfare of the public, and to disclose promptly factors that might endanger the public or the environment;

2. to avoid real or perceived conflicts of interest whenever possible, and to disclose them to affected parties when they do exist;

3. to be honest and realistic in stating claims or estimates based on available data;

4. to reject bribery in all its forms;

5. to improve the understanding of technology, its appropriate application, and potential consequences;

6. to maintain and improve our technical competence and to undertake technological tasks for others only if qualified by training or experience, or after full disclosure of pertinent limitations;

7. to seek, accept, and offer honest criticism of technical work, to acknowledge and correct errors, and to credit properly the contributions of others;

8. to treat fairly all persons regardless of such factors as race, religion, gender, disability, age, or national origin;

9. to avoid injuring others, their property, reputation, or employment by false or malicious action;

10. to assist colleagues and co-workers in their professional development and to support them in following this code of ethics.

Approved by the IEEE Board of Directors, August 1990

Figure 11-2   Ten ethical guidelines used by the IEEE. These also could be used to substantiate an ethical position you believe you must take.

achieve unlimited pleasure? To gain unlimited possessions or to live in harmony with nature? To live happily with one another or according to the dictates of a divine power? You are really the only one who can validly answer these questions for yourself, although others might have told you how *they* feel you *should* think or act. Whatever your personal outlook, it's worth remembering that the study of ethics will not necessarily make you a "better" person, but it will make you a more knowledgeable person when you come face to face with difficult professional decisions. We hope that this section has given you some insight and tools that will allow you to be an ethical researcher, writer, and engineer.

```
┌─────────────────────────────────────────────────────────────────────┐
│  ☐  What caused this dilemma in the first place?                      │
│  ☐  Have I clearly defined the dilemma and its possible options?      │
│  ☐  Should others be involved in any final decision?                  │
│  ☐  What are the immediate or long-term results of each option likely to be?  │
│  ☐  Could any option injure anyone (a) physically (b) emotionally (c) professionally?  │
│  ☐  Are all my options legal?                                         │
│  ☐  To what extent does each option follow the golden rule?           │
│  ☐  Will my decision be one I would willingly share with my           │
│         ☐  management?                                                 │
│         ☐  colleagues?                                                 │
│         ☐  family?                                                     │
│         ☐  lawyer?                                                     │
│         ☐  local news media?                                          │
│         ☐  religious leader?                                          │
│  ☐  Whatever option I choose, could there ever be exceptions to it?   │
└─────────────────────────────────────────────────────────────────────┘
```

Figure 11-3   Checklist for ethical decision making.

THE ETHICS OF HONEST RESEARCH

If a burglar made off with your stereo, you would know what to do. But what if a professional colleague stole your words? Plagiarism—the act of using someone else's work without giving proper credit—is a crime of intellectual property, and one might argue that it is just as serious as a crime of real property.

Pat Janowski, *The Institute* (IEEE), December 6, 2004.

The percentage of people who read this book and who would go to their neighbor's house and steal a laptop or stereo is extremely low (we hope). Yet a common kind of theft among students in high school and college is *plagiarism*. This kind of dishonesty is not limited to youth or the academic world. If as an engineer you knowingly or unknowingly "borrow" the language, ideas, or graphics of others, representing them as your own original work by failing to acknowledge your sources, you are plagiarizing—a very serious offense. You might even be infringing on someone's copyright, and thus could open yourself up to lawsuits.

Plagiarism is frequently the result of ignorance or carelessness rather than dishonesty. It's easier to replicate another's ideas than to put what you have read in your own words and then reference it properly. When you do research, *all information* that you obtain from journals, books, interviews, the Internet, or any other sources *must* be fully documented—that is, accompanied by references to the sources where you obtained the information. This includes all information, diagrams, ideas, facts, theories, findings, opinions, and graphics. And citing the sources of your information—"legal" plagiarism—lends greater authority and credibility to your documents. It shows you've done your homework.

The only exception to the rule of acknowledging your sources is when you cite common knowledge. But what is "common knowledge"? First, it's usually considered to be any fact, date, event, information, circuit, or equation that can easily be looked up in a standard reference book. However, what may be common knowledge to some may not be common knowledge to others.

Think of a generally accepted theory you learned in engineering school: You can find it in practically every standard textbook on the subject, and it's not documented when discussed in those books. That's common knowledge. But think of a new theory put forth by an engineer who is not well known in her field. That's not common knowledge, and if you use the theory in a report, you must say where you got it from—that is, you must document your source. The difference then comes down to your familiarity with your field, and whether you can distinguish what is common knowledge to your audience from information that is not.

A SYSTEM FOR DOCUMENTING YOUR SOURCES

When you "document" your information sources, you are indicating where you found your information and from whom you borrowed it. As an engineer, you are most likely to use the IEEE style of documenting your sources.

CITING INFORMATION

How you cite your sources depends on the documentation system you use (a standard system is provided below). Document your information borrowings in order to:

- Protect the originator, the author of the information, so that she or he will get the credit and acknowledgment for having developed it.
- Protect yourself from accusations of plagiarism—of stealing other people's hard-won discoveries.
- Demonstrate to readers that you have done your homework and are aware of the latest developments in the particular field.
- Enable readers to track down the information so that they can read it for themselves.

Note With the evolution of the World Wide Web, plagiarism is more prevalent than ever. You can access libraries, reports, journals, and graphics within seconds. This borrowed information too must be documented, including images or graphics from the Web. To counteract the rampant plagiarism that now takes place in colleges and universities, professors are currently using programs that can search the Web and find whether specific pages, paragraphs, and even sentences are stolen from the Web.

USING THE IEEE SYSTEM

The following pages give examples of how to document your sources. This information is based on the system used by the Institute of Electrical and Electronic Engineers (IEEE), but it is almost identical with systems used by other engineering organizations and industries. In the following, you will see an example of a well-documented page from an engineering student's report (Figure 11-4). Next, you will find guidelines on how to format a references page and an example (Figure 11-5) of a short page of references. Finally, you will see how to format the varied sources you might use in your research.

Procedures for Documenting

1. In the body of your text, refer to the source of your information by inserting source numbers in brackets at the end of each segment of borrowed information—like this [1]. This means that the borrowed information came from source 1 on your references page. Reference numbers can also be inserted within a sentence like this [2], without changing the sentence's punctuation. You can also cite your reference in your text thus: *According to the 2006 U.S. Census Bureau [3], consumption* ....

2. Unless you are referring to a book or article as a whole, identify the page number(s) of your source of information. Indicate exact page numbers of a source within your brackets after a comma [4, pp. 3–6], or by a simple rhetorical device in your text such as *However, on page 79 of [5] the author seems to contradict herself when she states* .... If you must refer to more than one source in the same reference, use semicolons for separation: [6, p. 46; 7, pp. 29–31; 9, pp. 8, 12].

3. References at the end of quotation marks are punctuated with the period after the reference, "like this" [8, p. 23]. Once you have numbered a source, use the same number for all subsequent references to that source throughout your work. Figure 11-4 shows a page from a research paper that is documented following the above specifications.

Reference Page Format

List your sources in numerical order according to when they are first cited in the text, *not* by alphabetical order of authors' last names.

Use the initials—*not* the full first names—of authors. Use sentence-style capitalization on the titles of journal articles and enclose them in quotation marks: for example, "Method of lattice quantification."

Single-space individual references, with following lines aligned with the first. Double-space between separate references.

6.0 THE FUTURE OF HEVs

Knowing exactly what the future holds for HEVs is impossible. However, using what we know to be true today, we can generally extrapolate to a reasonable degree what tomorrow might bring.

6.1 Options

With technology comes options, and hybrid technology is no different. There are many different ways in which a hybrid can be configured, and since each has its own advantages, many different options will most likely be offered to the consumer. "Rather than having only one propulsion system choice when buying a future vehicle, it may be possible to select the propulsion system in the same way that one selects a 4 cylinder engine or a V8" [10, p. 43]. One could choose from a conventional gasoline, battery only, or any number of configurations of an energy storage device and a hybrid power unit (HPU) [9, pp. 98-99].

6.2 Fuel cells

Though today's HEVs have a conventional gasoline or diesel engine combined with an electric motor, in the next five years we will most likely see the arrival of the fuel cell in hybrid vehicles [13, p. 11]. Much work—and money—is going into improving on this technology.

6.2.1 Brief overview of the fuel cell. Fuel cells generate electricity through an electrochemical reaction that combines hydrogen with air. Many different fuels can be used, but methanol is often the fuel of choice, with which the fuel cell's only emission is water vapor, making it the cleanest alternative available [1].

6.2.2 Current limitations of fuel cells. Unfortunately, fuel cells need further development in order for them to be feasible in personal automobiles. First of all, as with all new technology, the fuel cell is expensive. It will take some deflation of cost before it can match the cost of a conventional gasoline engine, and thus penetrate the market [16, pp. 14-16]. In addition, the fuel cell has not been a viable option due to its large size. However, great strides have been made in this area in the past few years, and "officials at DaimlerChrysler have pledged to have a viable, commercial fuel cell vehicle available in 2004" [16, p. 17].

In order to reform fuel (change it into its useful form so it can react to create energy), the system has to be heated to a certain temperature in order for the reaction to occur [13, p. 8]. Thus, long start-up times are also holding fuel cells back from use in HEVs, yet although there are still considerable strides to be taken in fuel cell technology, these cells will definitely serve as a viable option for HEVs in the near future [1].

6.3 Future models

Only two car companies have HEVs on the market today, but in the next few years almost all car companies are likely to follow suit [9]. As they flood the market, prices will drop, and the HEV will be cost comparable to a conventional vehicle. Below are some HEV models that might be emerging in the next few years.

6.3.1 Ford P2000 LSR. One model to be introduced shortly is the Ford P2000 LSR, which was delivered by the Ford Motor Company to the U.S. Energy Department in October, 1999. The P2000 LSR will be a hybrid diesel-electric vehicle with "the passenger room, trunk space, and driving acceleration of a Taurus" [17]. Ford has also designed the Ford Prodigy, a concept, diesel-electric hybrid family sedan that will get 80 miles to the gallon [18, p. 3].

8

Figure 11-4 A page from a well-documented research paper.

REFERENCES

[1] C.H. Roth, *Fundamentals of Logic Design*, 5ᵗʰ ed. St. Paul: West Publishing Company, 2003.

[2] R. Schneiderman, *Future Talk: The Changing Wireless Game*. New York: IEEE Press, 1997.

[3] N. Hart, "Mobile satellite system design," in M.J. Miller, ed., *Satellite Communications: Mobile and Fixed Services*, pp. 103–143. Boston: Kluwer Academic Publishers, 1993.

[4] K. Chang, "Surpassing nature, scientists bend light backwards," *The New York Times*, p. F4, Aug 12, 2008.

[5] *Catean Dinosuria Handbook*. San Diego: Elaine Research Corporation, 2005.

[6] Personal interview with Dr. Bill Fagelson, ECE Department, The University of Texas at Austin, November 18, 2007.

[7] L. Katayama, "Flame warrior," *Wired*, pp.110–117, June 2008.

[8] C. Hilary and D. Mor, "The power infrastructure," http://www.cs.dartmouth.edu/2K/power-CM/. Accessed April 2, 2001.

[9] C. Xiao, Y.R. Zheng, and N.C. Beaulieu, "Second-order statistical properties of the WSS Jakes' fading channel simulator," *IEEE Trans. on Communications*, vol. 50, no. 6, pp. 888–891, June 2002.

[10] Email from Mark A. Carpenter, A98-b2 project manager, AMD, Austin, Texas, March 8, 2008.

Figure 11-5 An example of a brief reference page. Note how spacing and alignment of each entry make for easy visual access.

Use a common abbreviation for a journal title if there is one, e.g., *IEEE Electron Device Lett.* Otherwise, give the full name of the journal.

End each entry with a period.

Even if you have referred to the same source more than once in your paper, list that source only once on your references page.

Drone aircraft violating privacy?

Not only are drones alleged to have mistakenly blown up Afghanistan weddings, but they can spot activity on the ground from thousands of feet in the air, act as a cell phone tower and get GPS info from cell phones, and thus pose a threat to the privacy of ordinary citizens.

For details, see the Preface for the URL.

SAMPLE REFERENCES

Following are examples of how items would be listed on a references page. They illustrate most of the kinds of references you will likely have to cite. If you have a source lacking a model for citation, provide enough information to allow readers to hunt down that source if they want to.

Book

[1] B. P. Lathi, *Linear Systems and Signals*. London: Oxford University Press, 2001.

Book, Multiple Authors

[2] S. Horner, T. Zimmerman, and S. Dragga, *Technical Marketing Communication*. New York: Longman, 2002.

New Edition of a Book

[3] C. Conrad and M. S. Poole, *Strategic Organizational Communication*, 5th ed. New York: Harcourt Press, 2002.

Journal Article

[4] N. M. Tahir, A. Hussain, S.A. Samad, and H. Husain, Shock graph for representation and modeling of posture, *ETRI Journal*, vol. 29, no. 4, pp. 507–514, August 2007.

Article in an Anthology

[5] G. J. Broadhead, Style in technical and scientific writing, in M. G. Moran and D. Journet, eds. *Research in Technical Communication: A Bibliographic Sourcebook*, pp. 379–401. Westport, CT: Greenwood Press, 1985.

Translation

[6] M. M. Botvinnik, *Computers in Chess: Solving Inexact Search Problems*. Translated by A. Brown. Berlin: Springer-Verlag, 1984.

Personal Interview/Communication

[7] Interview [or Personal Communication] with Prof. David Beer, ECE Department, The University of Texas at Austin, January 10, 2009. [Date omitted if unknown.]

Handbook/Data Book, No Author

[8] *Handbook of Accelerator Physics and Engineering*. Singapore: World Scientific Institute, 1999.

[9] *Engineering Ceramics Data Book*. New York: Engineering Materials Series, 1998.

[10] *ThinkPad T61 Service and Troubleshooting Guide*, 3rd ed. Morrisville, NC: Lenovo, 2007.

[11] Chelmsford, MA: Hittite Microwave Corporation, 2001.

Encyclopedia Entry

No author given:

[12] "Frequency," *Encyclopedia Britannica*, 2001 ed.

Author(s) given:

[13] D. G. Paxon, D. S. Wood, and W. C. Malden, Equity, in *The Blackwell Encyclopedia of Finance*, F. Carter, ed. Oxford, U.K.: Blackwell Publishing, Ltd. 1999. Online:

[14] Thermodynamics. *The New Online Britannica*, April 2002. http://search.eb.com

Course Notes

[15] M. Carpenter, *Lab Notes for EE464K, Senior Projects*. The University of Texas at Austin, Spring semester, 2008.

Dissertation or Thesis

[16] J. Kwan, *Internal Motivation in Classical Ethics*. M.S. Thesis, Plan II Honors Program, The University of Texas at Austin, 2007.

Proceedings Paper

[17] N. Coppola, "Computer-based training for chemists: Designing decision-making tools for green chemistry," in *Proceedings of the International Professional Communication Conference*, pp. 77–83, Portland, OR, Sept. 17–20, 2002.

Patent

[18] M. L. Chirinos, U.S. Patent 5 670 087, 2001. [Title of patent may be included.]

[19] M. Postol, "Method of lattice quantification which minimizes storage requirements and computational complexity," U.S. Patent 6 085 340, July 4, 2000.

Newspaper Article

[20] Virus overwhelms global internet systems, *The New York Times*, vol. 116, pp. A3, A8, January 27, 2003.

Government Publication

[21] *Basic Facts about Patents*. Washington, DC: U.S. Government Printing Office, 2002.

Technical Report

[22] R. Cox and J. S. Turner, "Project Zeus: Design of a broadband network and its application on a university campus." Washington Univ., Dept. of Comp. Sci., Technical Report WUCS-91-45, July 30, 1991.

[23] "TDDB results for 0.18 μm." Taiwan Semiconductor Manufacturing Co. Hsinchu, Taiwan, R.O.C., 2001.

Letter/Email

[24] Letter [or Email] from A. R. Hasan, Project Manager, Oracle, Boston, Massachusetts, January 5, 2007.

Software

[25] J. McAfee, *Virus Scan Version 6.0.* Computer software. Only available online. Networks Associates Technology, Inc. IBM-PC, 2001.

Database/Online

[26] R. Berdan and M. Garcia, *Discourse-Sensitive Measurement of Language Development in Bilingual Children.* Los Alamitos, CA: National Center for Bilingual Research, 1982. (ERIC ED 234 636).

[27] J. Ozer, External solutions for your expanding video library, *PC Magazine*, Jan. 27, 2003, v22, n10, p. 247(7) in Academic Index (database on UTCAT PLUS system).

World Wide Web

[28] "AT&T enters Indiana residential local phone market," www.att.com. Accessed January 26, 2003.

[29] "Nokia introduces the world's first handset for WCDMA and GSM networks," http://press.nokia.com/pr20023.html. Accessed January 27, 2003.

[30] B. L. Evans, "Brian Evans' home page," www.ece.utexas.edu/~bevans. Accessed February 12, 2003.

Slides and Films

[31] L. J. Mihalyi, *Landscapes of Zambai, Central Africa.* Santa Barbara, CA: Visual Education, 1975. (slides)

[32] *An Incident in Tiananmen Square*, 16 mm, 25 min. San Francisco: Gate of Heaven Films, 1990. (film)

Videocassette/DVD

[33] *Behind the Lines.* 96 min. Santa Monica, CA: Artisan Entertainment, 1997. (videocassette)

[34] F. W. McMaster, *Matrix Algebra for Electronic Circuit Analysis.* Flower Station, Ontario: Cottage Publishing. (video instruction tape) No date.

[35] *The Standard Deviants*: *Physics, Part 2.* Lorton VA: Cerebellum Corp., 1999. (DVD)

[36] *The Great War: Story of World War 1: Parts I & II.* London (UK): Eagle Rock Entertainment Ltd., 2005. (DVD)

> **Keystone pipeline to pollute the Great Plains?**
>
> The 1,700-mile, $7 billion pipeline that would transport crude oil from Canada to refineries in Texas has sparked great controversy and vehement opposition, particularly in Nebraska. Cozy relationships and a compromised environmental report are alleged.
>
> For details, see the Preface for the URL.

EXERCISES

1. On the Web, search "codes of ethics for engineers." Compare them. Think of situations in your career where you would be glad to have such codes to support you in your actions.

2. Access the Murdough Center for Engineering Professionalism at www.niee.org. In particular, explore the case studies of actual engineering problems. How can a center like Murdough be important to the profession?

3. Ask any engineers if they have had to make ethical decisions in their career. What was the nature of the dilemma? How did they make their decision? Were there any repercussions? Were they happy with their decision?

4. Research the disasters of the Challenger and Columbia space shuttles. What ethical questions did these disasters raise, both from a human and an engineering perspective? Perhaps, investigate other well-known tragedies in the automobile, nuclear, shipping, construction, or other industries. What do they illustrate about ethics and engineering decisions?

BIBLIOGRAPHY

ENGINEERING ETHICS

Davis, Michael. Thinking Like an Engineer: The Place of a Code of Ethics in the Practice of a Profession, in *Philosophy & Public Affairs*, *20*(2) (Spring 1991): 150–167.

Fleddermann, Charles. *Engineering Ethics*, 3rd ed. Upper Saddle River, NJ: Prentice Hall, 2007.

Landis, Raymond B. *Studying Engineering: A Road Map to a Rewarding Career*, 2nd ed. Los Angeles: Discovery Press, 2000.

Martin, Mike W., and Roland Schinzinger. *Ethics in Engineering*, 4th ed. New York: McGraw-Hill, 2004.

Murdough Center for Engineering Professionalism. Texas Tech University, www.niee.org, 2008.

Reynolds, George. *Ethics in Information Technology*, 2nd ed. Florence, Kentucky: Course Technology, 2006.

CITING INFORMATION*

The Chicago Manual of Style, 15th ed. Chicago: University of Chicago Press, 2003.

Engineering.com. Engineering Ethics: Case Studies. www.engineering.com/Library/ArticlesPage/tabid/85/articleType/CategoryView/categoryId/7/Ethics-Case-Studies.aspx. Accessed July 3, 2012.

Jones, Dan, and Karen Lane. *Technical Communication: Strategies for College and the Workplace*. New York: Longman, 2002. See particularly "Appendix C: Documentation Styles."

Radford, Marie, Susan Barnes, and Linda Barr. *Web Research: Selecting, Evaluating, and Citing*, 2nd ed. Boston: Allyn & Bacon, 2005.

Ruszkiewicz, John, Maxine Hairston, and Christy Friend. *SF Express*, New York: Pearson Education, 2002.

*Several websites give guidelines and examples for citing sources in various engineering disciplines. Many such sites can be found by putting "Citing Engineering Sources" in your search program.

ENGINEERING YOUR ONLINE REPUTATION

Regard your good name as the richest jewel you can possibly be possessed of—for credit is like fire; when once you have kindled it you may easily preserve it, but if you once extinguish it, you will find it an arduous task to rekindle it again. The way to a good reputation is to endeavor to be what you desire to appear.

—Socrates (circa 469 BCE–399 BCE), Greek Philosopher

INTRODUCTION TO SOCIAL MEDIA MANAGEMENT

Can you remember a time when we did not have the Internet? Some of us can remember when we had to rely upon the story or message that a business placed in its newspaper advertisement. We may have seen a TV ad that embellished the "facts" about a product, but we had no way of knowing if those words were true.

Have you ever stood at the fax machine and impatiently waited for your 12 pages to go through successfully? What about playing phone tag with an integral part of your engineering team? Not too long ago, engineers had to fly to meetings, stay in hotels, put up with unintelligible conference calling systems, be away from their families, and hope that the important people made it to the same meeting.

In 2012, the Mckinsey Global Institute released a study entitled *Unleashing Value and Productivity Through Social Technologies*, where it is estimated that social media could add $1.3 trillion to the economy in the next six years. Of special note were the fields of automotive, mechanical, and aerospace engineering. The author of the article,

Quentin Hardy, states that "since they work with a lot of autonomy, but also in consultation with others, [engineers] benefit the most from knowing such things as which employees have the deepest knowledge in certain subjects, or who last contributed to a project, and how to get in touch with them quickly."[1]

Typically, engineering positions are a collaborative-type of work environment and their communication efforts do not function well using the "one-way street" model anymore. Now that social interaction venues are prevalent in the workplace, there is no turning back to the old way of thinking—relying on word-of-mouth or believing the company line. Employees and employers are diligently working proactively to be informed, stay informed, and share vitally important data with others in worldwide collaborations that bring startling results.

Of course, TV ads still scream as we passively sit on our couches and soak in the information. And, consumers still largely base their opinions on the story that they are told by whoever paid for the newspaper, television, magazine, or radio advertisement. Some of that one-way information model will never be obsolete.

Historically, employees believed what they were told, and interacting with other divisions within the same organization, or affiliate organizations was seen as somewhat traitorous. Social media platforms change all that. "The proper use of social media tools adds to productivity, an improved consumer focus as well as better-functioning teams. Data and knowledge are exposed and shared instead of being hoarded," says Michael Chui, one of the authors of the McKinsey report.

ENTERING THE INTERNET AGE

The Internet has brought a virtual tidal wave of new possibilities, ideas, and methods that engineers can use for communicating to the world and has blown the one-way communication model to smithereens. The most profound change is that social media, and the messages contained therein, are not owned by any business or organization. Social media are owned by all of us. Engineers collaborating on one project, from all over the world or in the next cubicle is what it is all about!

In her book, *The Zen of Social Media Marketing*, Shama Kabani states that social media consists of "multiple online mediums all controlled by the people participating within them—people who are busy having conversations, sharing resources, and forming their own communities. Unlike radio, television, and print, it isn't passive—users don't just receive content; they create it, too."[2]

There is one element that historical marketing and social media campaigns have in common. Both methods are designed to get consumers to *take some kind of action*. Businesses still design and implement advertising campaigns to encourage consumers to take action. The Internet provides a broader platform for engineering businesses and

---

[1] *Source:* http://bits.blogs.nytimes.com/2012/07/25/mckinsey-says-social-media-adds-1-3-trillion-to-the-economy
[2] Kabani, Shama, *The Zen of Social Media Marketing*, Dallas, TX: Ben Bella Books, 2012.

organizations to promote their wares by leveraging a well-defined, strategically planned, and open communication style that encourages online communities (composed of consumers) to contribute to the message and branding of the product or service.

What the Internet and social media channels bring to the old marketing equation is that consumers have stepped into the massively important role of using their own voice to encourage or *discourage* other consumers to take action. Consumers use their positive or negative experiences to influence other consumers; this process is a powerful component of social media communities.

CONQUERING YOUR FEARS OF SOCIAL MEDIA

Social media has become a source of fear and confusion for many. As with anything new, there are fear-mongers and critics. There are also champions and evangelists. This new method of communication can be wildly successful for engineering businesses, or it can be an abysmal failure.

This chapter should dispel some of the confusion associated with using social media to promote yourself or your business. You will find a social media venue that fits your needs. Discussed in the following pages are five of the most widely used social venues, including WordPress blogs, Facebook, Twitter, LinkedIn, and Google+.

Some experts believe that having no online presence is tantamount to having a *bad* online reputation. Have you ever read a bad review of a restaurant online and then decided not to go? What would you do if that restaurant's manager had responded to the bad review and helped the disgruntled customer? Would you give the restaurant a second chance? Most of us would. This is one way that social media can benefit businesses.

But social media is not just for entertainment. *Information Week Security* performed a survey of 1,153 engineers in 2010 and found that, "of those who work in the aerospace, automotive, and commercial-vehicle industries, 55% of respondents said they use social media sites for work-related reasons, and that over half are allowed access while at work."[3]

Social media is not a get-rich-quick scheme, and it does not transform your life or engineering business overnight. It is, however, necessary for success in today's digital world.

> ### *An umbrella that charges a smart phone?*
>
> Vodafone's Booster Brolly is an umbrella that uses on-board solar cells and a micro antenna to charge your phone and boost its 3G signal. Engineering students at University College of London worked with Vodafone to develop this cool technology.
>
> For details, see the Preface for the URL.

[3] "Engineers Use Social Media for Business." Information Week Security. www.informationweek.com/news/security/management/225900054

CREATING A WORDPRESS BLOG

Chances are that you have read an engineering blog within the past month. You may have visited that blog on purpose, you may have stumbled upon it by accident, or you may have landed on the blog and never knew that it was a "blog."

UNDERSTANDING WHAT A BLOG IS

A "web log," or "blog" for short, is a specific type of website that is interactive. The site is *supposed* to be updated frequently with new articles and information about specific topics. Blogs contain information from lots of different sources. For example, blogs' owners (called "bloggers") can write articles themselves, or they can post articles that they found on other websites. The information can also originate in print, or it can be contributed by the readers of the blog.

That is it. There is nothing mysterious about a blog. There are CEO blogs, celebrity blogs, food blogs, travel blogs, graphic design blogs, architecture blogs and *engineering blogs*. Actually, there are too many types of blogs to list. The most important thing to remember about a blog is that it is interactive and should be constantly changing. This characteristic benefits both the reader and the blogger.

PARTICIPATING IN A BLOG

By viewing and participating in a blog, engineers can be educated, drawn in to a worthwhile conversation, join an online community of like-minded people, help build a branding strategy, shape ideas by giving their opinion, or give advice to product designers. Therefore, readers benefit by receiving information, ideas, or the chance to get their opinion out there.

No matter what the reason is for creating the blog, the blog's creator must remember that the blog reader is a partner in the success of that blog. Perhaps some bloggers want to become engineering experts in the "Bending Moment." If that is the case, their blogs will reflect their expertise in the elements involved in the Bending Moment. Those bloggers would use titles and searchable keywords to funnel traffic to their blog about the Bending Moment. This will allow the engineering blog to be found by those who are searching for this topic.

USING A BLOG FOR PROFESSIONAL ADVANTAGE

Engineers often use their blogs as a means of collaboration and networking within their field. Blogs are an excellent platform to give and receive advice and expertise from others engineers in your field. Who knows—you may end up getting an engineering job from a blogging contact! For example, a small group of engineers has created a blog and the tagline for their site is "Engineers writing about the workplace, common engineering problems, and lessons to share with the world."[4]

Here is an example of using a blog to build a reputation as an engineering expert:

1. Blogger A (an engineer) needs advice from experts about a faulty modeling process. He finds several blogs that discuss this topic, and he poses some questions on the blogs. By posting his question on a well-known blog, he increases the visibility of his own blog.

2. Blogger B, an engineering expert in the area of faulty modeling processes, sees the post (by Blogger A). Because she is trying to build her online reputation as an expert in faulty modeling processes, she posts a reply to Blogger A's question. She has just "advertised" to the world that she is an expert in faulty modeling processes.

3. After several online (and highly visible) conversations, Blogger A takes the advice of Blogger B. He follows the steps advised by Blogger B, and the issue with the faulty modeling process is resolved!

4. Blogger A thanks Blogger B (again, online for all to see) for the great advice and tells her that her advice resolved his issue. Blogger A looks good to his clients or employer because he resolved the faulty modeling issue.

5. Blogger B has furthered her efforts at building her reputation as an expert in faulty modeling processes. Her advice contributed to a real-life project.

6. Six months later, a person (prospective new client) conducts a search about "faulty modeling process." The related engineering blog comes up in the search because Blogger A's and Blogger B's posts used searchable keywords. The person sees both sides of the conversation on the blog and sees that Blogger B resolved the issue for Blogger A. The person sees Blogger B as an *expert* and contacts her to help resolve a similar issue.

7. Blogger B scores some new business because of her online reputation!

In the above example, all the engineers win via the use of this blog. Blogging brings people together even if they live in three different countries and never meet each other in person.

See www.electricalengineeringonline.net/engineering-blogs for a list of the 50 top *engineering blogs* found on the Internet.

Choosing Your Blog Software

Engineering students and professionals alike can benefit from creating their own blogs. You might think that the example about Blogger A and B was make-believe. Fortunately for you and other professional engineers throughout the world, the scenario happens every day. The sky is the limit when it comes to the usefulness of blogging for engineers.

---

[4]*Source:* www.engineeringblogs.org

Currently, the most user-friendly and intuitive blogging software available is WordPress. Visit its site at: http://wordpress.org

WordPress has some excellent characteristics:

- It can be used to set up a free blog, although you can set up a custom URL for as little as $17 per year.
- It does not require a single bit of programming knowledge.
- It comes with countless tutorials and technical support.
- It can be used for e-commerce by adding a shopping cart just in case you ever decide to monetize all of your expertise!
- It can be updated or reorganized easily by a complete novice.
- It offers almost limitless graphic themes to promote yourself as an individual or for your company brand, including custom colors, layouts, widgets, and graphics.
- It provides constant updates for free.

EXPLORING ENGINEERING BLOGS

If you are to understand blogging, you will have to visit a few. To find blogs that are relevant to engineering, it is as simple as typing "engineering blog" into any search field.

There are websites that rank the best engineering blogs on the Internet, too. That is the best place to start your search. For example, type the following in to your browser: www.invesp.com/blog-rank/Engineering

Once you see the list of the top 25 engineering blogs, choose several to view. While you are viewing each blog, consider the following:

- Is the blog visually attractive or distracting?
- Is the content relevant and up-to-date?
- Is it easy to see what specific areas are discussed on this blog?
- Is there an easy way to interact with the blog's creator?
- If needed, does the blog provide evidence of professional engineering clout?

After reading a few blogs, you will begin to understand the differences in style, content, and professionalism. You will also see the big difference between a traditional information-only website and the interactivity of a blog.

Remember, the main point to keep in mind is that a blog gives you the ability to interact with the blogger and others who view the blog. When thinking about designing your blog, make sure that the conversational style, community feel, and content are all open and inviting.

> **Turning body heat into energy?**
>
> Now there is a new source of power that could be put to use; human body heat. Engineers in Stockholm are the first to use the body heat of 250,000 daily commuters in one train station to heat a building next door. Wow.
>
> For details, see the Preface for the URL.

You can visit my blog at: http://jillbrockmann.com

What will you write about on your blog? Visit the companion website to see step-by-step instructions for designing and implementing your own blog. (See the Preface for the URL.)

BUILDING A FACEBOOK PAGE FOR A BUSINESS

If you have never heard of Facebook, then you have probably been living under a rock.

Here are some factoids about Facebook's worldwide reach:

- Facebook is used by 70% of automotive and aerospace engineers to exchange information on technical issues, according to a survey of SAE International.[5]
- Facebook had over one billion monthly active users at the end of December 2012.
- On average, it had 526 million daily active users.
- In March 2012, there were 488 million monthly active users who used Facebook mobile applications.
- Facebook had more than 600 million mobile monthly active users as of December 5, 2012.
- During December 2012, on average 500 million users were active with Facebook on at least six out of the last seven days.[6]

HARNESSING THE POWER OF FACEBOOK

If you were offered a customized advertising campaign for your engineering business—and it promised to reach millions of people, worldwide for *free*—would you bite? What if you were offered unlimited access to a panel of 10,000 civil engineering experts (from all over the world) that could help you unravel an issue with a traffic pattern—again, all for *free*?

Consider this:

Did you know that every single person who interacts with your company or organization via social media inadvertently becomes your cheerleader or critic? You may have heard the old adage that if you like a business, you will tell three people. If you *don't* like it, you will tell ten people.

This phenomenon is what makes applications like Facebook such great venues for engineers. Originally, Facebook was created to allow college students to connect

---

[5] *Source:* www.informationweek.com/news/security/management/225900054
[6] Facebook statistics, http://newsroom.fb.com/content/default.aspx?NewsAreaId=22

and share ideas, stories, events, and generally "socialize." Now, people from all over the world and from all professions use Facebook in order to share information, be heard, and make connections. Instead of verbally telling ten people they didn't like a business, they are telling millions of people via Facebook. Instead of having one or two aeronautical engineering experts to consult with, there are now thousands of them at their fingertips.

For example, a Facebook page called "Interesting Engineering" has almost half a million followers. Figure 12-1 shows a screenshot of what the home page looks like. You can find this site by typing in the following URL in your browser: www.facebook.com/interestingengineering

Interesting Engineering's Facebook page says, "If engineering is a headache for you, join us. We will change the way you feel."[7] How cool is that? They are using Facebook to demystify engineering and target a younger audience.

As of this writing, the Facebook page in Figure 12-1 has 489,000 individual "likers," and 337,060 people are talking about the content (a "liker" is an individual that has subscribed to the news posted on this Facebook page). This is an example of how Facebook creates a community of like-minded people who share an interest in engineering, opinions, and ideas. The people in this community can connect, share ideas, vent their frustrations, seek advice, talk about their successes, post project ideas, and discuss all kinds of engineering-related subjects.

Figure 12-1   Facebook business page or fan page named "Interesting Engineering."
Source: www.facebook.com/interestingengineering

---

[7]*Source:* www.facebook.com/interestingengineering/info

Here is an example of how engineering students and professionals can harness the power of Facebook:

1. An engineering student is lurking around a classmate's Facebook page and sees that her classmate has "liked" a page titled "Interesting Engineering." She decides to click the live link and visit that page.

2. As she scrolls through the engineering-related articles and posts, she stumbles across an article that discusses the subject of her thesis, wind turbine grid development.

3. The student reads the article and scrolls through the reader comments posted. One of the participants, an engineering expert in the topic, has an unusual opinion about wind turbines. The student clicks on the expert's name and immediately lands on the expert's Facebook page.

4. The expert's Facebook page has an About section that lists an email address, phone number, website, blog address, professional associations, education details, and current employer information. The student emails the expert to ask for a telephone interview in order to get some quotes for her thesis.

5. The expert writes back, agrees to the interview, and schedules it for two days hence.

6. Two days later, she gets her interview and obtains valuable information from a well-known engineering industry expert. It adds immeasurably to the content of her thesis.

7. The expert is so impressed with the passion and insight of the student, he offers her an internship that turns in to a full-time engineering position after she graduates.

In the above example, both the engineering student and the expert benefit from the connection they made via Facebook. This type of collaborative connection and sharing happens every day on Facebook. Let us not forget that all of this interaction, conversation, learning, and community building happens all across the world, seven days a week, 24-hours per day—for *free*.

USING FACEBOOK TO BUILD YOUR BRAND

You can use three methods to create your online presence (also referred to as your "brand") and display information on Facebook. Luckily, Facebook provides step-by-step instructions to help you create profiles that can be customized to spotlight engineering professionals.

Note   Remember that every friend made, page liked, photograph posted, group joined, event created, or comment left is a reflection of who we are or who we want the world to perceive us as being. If you are creating your social media presence to reflect your engineering business, choose your moves carefully. Visit the companion website to see step-by-step instructions for designing and implementing your customized Facebook page.

The three ways to create a profile on Facebook that presents your engineering business are as follows:

1. **Individual profile.** This is referred to as the "human being" profile. It is for an individual person, *not* a business. Anyone with an email address can build one of these pages. An example of an individual Facebook page is shown below in Figure 12-2.

2. **Fan page.** This is also referred to as a "business page": it could be for a product, service, organization, club, politician, or any other business-related entity. In order to create this page, you have to have an *individual profile* first. See Figures 12-3 and 12-4 for an example of a fan page for the engineering company, Structures, PE, LLP.

   A Facebook fan page comes with some useful features for businesses. One feature is a "People" page. This allows businesses to create a page that shows off the expertise or specialization of their employees. Figure 12-4 shows an example of the engineers at Structures PE, LLP.

3. **Individual fan page.** This type of page is a hybrid. It is just like the "business page," but it is used for an individual who is an engineering professional. In order to create this page, you have to have an *individual profile* first.

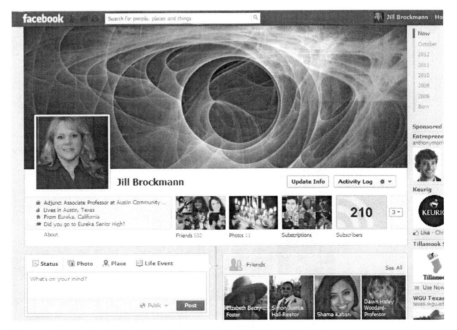

Figure 12-2   Individual Facebook page for Jill Brockmann.
Source: www.facebook.com/jillbrockmann

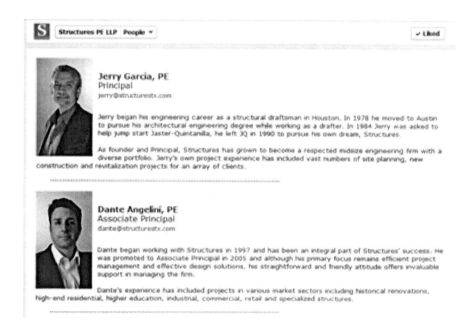

Figure 12-3   Structures Fan Page on Facebook.
Source: www.facebook.com/Structures, PE, LLP.

Structures PE LLP   People ▾   ✓ Liked

Jerry Garcia, PE
Principal
jerry@structurestx.com

Jerry began his engineering career as a structural draftsman in Houston. In 1978 he moved to Austin to pursue his architectural engineering degree while working as a drafter. In 1984 Jerry was asked to help jump start Jaster-Quintanilla, he left JQ in 1990 to pursue his own dream, Structures.

As founder and Principal, Structures has grown to become a respected midsize engineering firm with a diverse portfolio. Jerry's own project experience has included vast numbers of site planning, new construction and revitalization projects for an array of clients.

Dante Angelini, PE
Associate Principal
dante@structurestx.com

Dante began working with Structures in 1997 and has been an integral part of Structures' success. He was promoted to Associate Principal in 2005 and although his primary focus remains efficient project management and effective design solutions, his straightforward and friendly attitude offers invaluable support in managing the firm.

Dante's experience has included projects in various market sectors including historical renovations, high-end residential, higher education, industrial, commercial, retail and specialized structures.

Figure 12-4   Structures PE, LLPs People page on Facebook.
Source: www.facebook.com/StructuresPE/app_7146470109

For example, an individual engineer may have two Facebook pages:

- One page for his or her *private life*. This *individual* page includes family, friends, hobbies, political views, music preferences, jokes, religious affiliations, and vacation photos. This page is fun, relaxed, family-oriented, personal, and private.
- The other page lists the person's name as the *business name*, and it contains professional and engineering-related information only. This *fan page* is created to promote the person's engineering expertise or specializations. This page contains no personal attributes that could be controversial such as political or religious views. This page is strictly a method of promoting one's engineering expertise on Facebook.

The advantage of creating a Facebook fan page is that you can see demographic data about the people visiting your page. You do this by analyzing Facebook traffic using a tool called "Insights." It enables you as the page owner to plan successful content and interactions. Again, Facebook provides all of this data free of charge to business (fan) pages. Visit the companion website to see instructions about how you can use the Insights data to make your Facebook fan page more attractive to readers. (See the Preface for the URL.)

The image shown in Figure 12-5 is an example of what one of the Insights pages looks like.

Figure 12-5 Facebook Insights demographic page for Business and Technical Communication at Austin Community College.
Source: www.facebook.com/BTCMACC

When creating and building your brand via a Facebook page, consider the following:

- Will you create a personal page and a professional page?
- How can you spotlight your field of engineering?
- Can you visualize how you could use the commenting functions available on Facebook to gain credibility and become known as an expert in your engineering specialization? For example, when you comment on articles found on Interesting Engineering's Facebook page, you will be seen by half a million people.

> **Want to live forever?**
> **Wait until 2045.**
>
> Dmitry Itskov, a wealthy Russian, is creating "Avatar." First, he will transplant human brains into robots and then reverse-engineer the brain to effectively "download" human consciousness onto a computer chip. He estimates that by 2045 humans can buy immortality.
>
> For details, see the Preface for the URL.

- How will you make the design of your Facebook page look consistent with your blog, website and other social media platforms?
- Subscribe to me on Facebook: www.facebook.com/jillbrockmann for tips and tricks about using Facebook pages to promote your personal brand online.

Visit the companion website to see step-by-step instructions for designing and implementing your very own Facebook pages. (See the Preface for the URL.)

USING TWITTER TO CONNECT AND SHARE INFORMATION

If you have never heard of Twitter, then you have probably been living under another rock.

Here are some interesting facts about Twitter:

- Many engineering companies, large and small, now use Twitter to communicate with existing and potential customers. Also, many professional engineering publications post the latest news and links to their online articles on Twitter.[8]
- Even though it is a young social media venue, it has 500 million registered users daily, and 33 billion tweets sent worldwide—*daily*.[9]
- Anyone can read, write, share, or re-share messages about a field of interest, as long as the message is 140 characters or less.

---

[8] *Source:* http://blog.prosig.com/2011/02/24/how-can-social-networks-help-engineers/

[9] Twitter Facts and Stats, by Bran Friedman, March 14, 2012. *Source:* http://socialmediatoday.com /bradfriedman/469107/twitter-facts-and-stats

- "Tweets," or typed messages, are posted in real-time and can be read by anyone, from anywhere, regardless of whether they are logged in to their Twitter account.
- Twitter is fast becoming the "go to" medium for engineering information, according to Bosch Rexroth, cited below.

After launching a pilot project to study how engineers used social media, Kevin Gingerich from Bosch Rexroth (see Figure 12-6) stated, "We were astonished by the vibrancy of the engineering and industry communities on *Twitter*. It's become our engine of choice for communicating to a broad audience, from recent innovations and upcoming exhibitions to new technical references on our website."[10]

USING TWITTER TO CONNECT WITH OTHER ENGINEERS

Why should you or your business join those who swear by Twitter for connecting to others in the engineering field? Here are a few reasons:

- If you are interested in spreading the word about your own engineering expertise or the specialty of your company, why would you ignore a virtual database of opinions, trends, messages, stories, and networking opportunities?
- In the engineering context, Twitter is an environment that is rich with ideas, conversations, and content from influential industry leaders. Don't forget that

Figure 12-6   Twitter account home page for Bosch Rexroth Corporation.
Source: https://twitter.com/boschrexrothus

---

[10]Machine Design.com. Twitter for Engineers. *Source:* http://machinedesign.com/article/twitter-for -engineers-can-social-media-play-a-role-in-the-design-community-0302

all of this sharing, communicating, and brand-building activity is available 365 days a year, 24 hours per day—for *free*.

Figure 12-6 is an image of a successful engineering home page on Twitter.

Businesses can use Twitter # (hashtags) to corral participants at a company-sponsored event, for curating crowd-sourced data about product research, and for connecting directly with engineering clients and customers. More information about the importance of hashtags and crowd-sourcing data can be found on the companion website. See the Preface for the URL.

- Engineering professionals can use Twitter to funnel traffic to special events, online design contests, interactive websites, new blog entries, product launches, and other information-rich connections with consumers.
- Individual engineers can use Twitter to foster strong networking relationships with other engineers by listening and gathering industry intelligence about competition, market trends, and employment opportunities.

Imagine you are a graduating engineering student; here's how you can use Twitter:

1. You are about to graduate from a small college with a mechanical engineering degree and have arrived at the International Manufacturing Technology Show (IMTS) 2012 in Chicago. You are to visit the booth of the internationally known company that paid for your flight, hotel, and admission to this event. You have been told that it is located in Booth #10602. Only, they are not actually at that location. No other information is available. Hmm. What to do?

2. Enormous Engineering Company (EEC), the company that paid for your dream trip, had to move its booth to a different building because of mechanical "issues" with Booth #10602. EEC is worried that attendees will not be aware of its last-minute location change. The company tweets immediately about its new building location at IMTS.

3. You are following EEC on Twitter. You quickly scroll through your tweet feed to see if there is any word from them. Bingo! They have moved to a different building three blocks away.

4. You inspire and amaze EEC's booth members by arriving early to express your hearty thanks for the great trip, and you make sure that their mechanical issue is fixed at the new location. They are grateful for your arrival at the new location and glad that Twitter helped route you to the new location.

5. Upon returning home from your trip you find a letter from the president of EEC, thanking you for your help with the booth and inviting you to dinner to talk about your plans for after graduation.

In the above example, Twitter moved mountains for you and EEC. When a company invests thousands of dollars setting up plans for an enormous event like IMTS, it can be derailed quickly by a sudden event like changing a booth location. Since Twitter is known for instantaneous communication, it is a perfect platform for this occurrence. EEC could have missed a few thousand booth visitors, and you may not have found them to say "thank you" for your trip. You would have missed the free dinner invitation, too.

CUSTOMIZING YOUR TWITTER ACCOUNT

When creating the account for you or your engineering organization on Twitter, consider the following:

- Make sure others can find you easily on Twitter. Customize your Twitter user name, profile photo and design, bio, and online appearance by maintaining brand consistency with colors, images, and logos. Use keywords in your bio to help people find you.
- Don't worry about how many followers *you* have. Instead, concentrate on following engineering industry leaders, cutting-edge companies, and experts in engineering. Read their tweets, place comments, re-tweet them, and actively seek information that will bring you new information and ideas. Your own followers will come eventually.
- Use customized searches and hashtags to filter out everything except the information you need.
- Attach hashtags to photos or videos you upload so they are easily found in searches. Remember to use authentic keywords whenever possible.
- To find people, companies or engineering organizations to follow, use **Twellow.com** and **WeFollow.com** directories on Twitter that are categorized by interest and industry.

Follow me on Twitter: @JillBrockmann

Who will you follow on Twitter? How will you manage the information and communications you receive on Twitter? To get answers to these questions and others, visit the companion website to see step-by-step instructions for designing and implementing your very own Twitter account. (See the Preface for the URL.)

> ### Does adding sound to cars increase safety?
>
> Engineers have developed a vehicle warning system that improves safety for bicyclists. The system consists of a GPS-enabled device mounted to the dashboard of an electric car (which are dangerously quiet). It warns bicyclists of the approach of a quiet car.
>
> For details, see the Preface for the URL.

GENERATING YOUR INTERACTIVE RÉSUMÉ ON LINKEDIN

In a recent *Wall Street Journal* article titled "No More Résumés," Rachel Emma Silverman stated that "Instead of asking for résumés, a New York venture-capital firm—which has invested in Twitter—asked applicants to send links representing their 'Web presence.'"[11] What? No résumé?

---

[11] Rachel Emma Silverman, "No More Résumés," *Wall Street Journal*, January 24, 2012. *Source:* http://online.wsj.com/article/SB10001424052970203750404577173031991814896.html

Tech-savvy employers are acquiring evidence of people's web reputation and searching for their online interactions as a means of finding better-quality candidates —especially for engineering companies or organizations that rely heavily in the Internet and social media to build their brand, enhance employee collaboration, and reach their customer base.

INCREASING VISIBILITY FOR ENGINEERS

When it comes to actively participating in various social media platforms, the Society of Automotive Engineers International (SAEI) has determined that "engineers are apparently just as hip as the rest of us." SAEI has drawn this conclusion after completing a survey that revealed 61% of engineers polled use social media sites like Facebook, LinkedIn, Twitter, and YouTube.[12]

LinkedIn is the perfect place to increase online visibility, showcase your engineering expertise, and network with other professionals in your field. Even though LinkedIn is considered a "social" media site, the purpose of creating and maintaining a profile on this platform is solely for professional networking, connecting, and building career-oriented associations.

Figure 12-7 shows a typical LinkedIn profile of engineer, author, and speaker Anthony Fasano, PE, LEED, AP.

Completing your LinkedIn profile can be time-consuming, but the benefits are enormous. The tools provided allow space to include all phases of your professional life including your work experience, education, business associations, personal and company websites, other social media accounts, volunteer activities, areas of special expertise using keywords, and other ways to enhance your professional message to other LinkedIn members.

CREATING YOUR LINKEDIN PROFILE

When creating your profile on LinkedIn, consider the following:

- **Complete your entire profile.** A sparsely completed profile is almost as bad as having no profile at all. Your profile should be compelling, interesting, accurate, and complete. Include as much engineering-related information as possible. Remember to use keywords.

- **Actively participate in LinkedIn Answers.** The answers feature helps nurture professional engineering connections and gives you a chance to show your expertise by answering questions posed by others. Participate by posing questions yourself, too.

- **Join groups.** Find groups related to the engineering profession, join them, and actively participate in ongoing conversations to get your name seen in the forums. If there are not any groups about your specific area of engineering, create one!

---

[12]"Mobility Engineers Big Users of Social Media." Society of Automotive Engineers International. *Source:* WWW.SAE.ORG/JSP/JSPS/MKTWHITEPAPERFORM.JSP?PUBL=SOCIAL

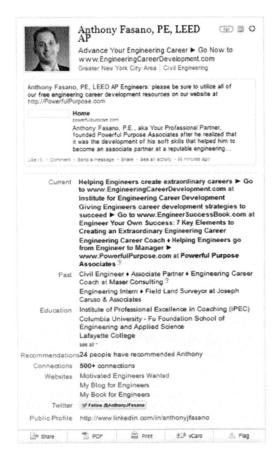

Figure 12-7 LinkedIn profile home page for Anthony Fasano, PE, LEED, AP.
Source: www.linkedin.com/in/anthonyjfasano

- **Use Advanced Search.** The basic search function is great, but Advanced Search has all sorts of ways to fine-tune your results, including engineering keywords and product names in all your posts.
- **Treat your LinkedIn profile similar to a website.** Make sure your profile is up-to-date, organized, well-formatted and contains lots of interesting engineering-related information. It goes without saying that misspellings are a no-no!
- **Populate your profile with keywords.** Use keywords that reflect your background, education, professional field, or expertise. Use variations of those words, too: for example, use engineer, engineering, engineered.
- **Increase the chances you will be found with search engines.** Make sure that your profile is marked as "public" so that you will show up in searches for your profession.

GIVING AND RECEIVING RECOMMENDATIONS

One of the best features offered to LinkedIn members is the Recommendation section. This is a profile portion designed to highlight your achievements and the exceptional work of other LinkedIn members.

You may politely ask an engineering colleague for a recommendation for any collaborative work you performed. You may receive a request for a recommendation from another LinkedIn member.

People in charge of making hiring decisions commonly review LinkedIn profile pages. It is important that you request recommendations that are relevant to your engineering specialty. Make the most of the section by requesting recommendations from colleagues you have collaborated with and by giving accurate and detailed recommendations to those colleagues you feel should be spotlighted. Figure 12-8 shows an example of the engineering recommendations section on the home page of an engineer's LinkedIn profile.

> ### First algae-powered car?
>
> The world's first algae fuel-powered vehicle, dubbed the Algaeus, was revealed in San Francisco. The plug-in hybrid car, which is a Prius equipped with a nickel metal hydride battery, runs on green algae.
>
> For details, see the Preface for the URL.

Recommendations For Anthony

Engineering Career Coach ♦ Helping Engineers go from Engineer to Manager ► www.PowerfulPurpose.com
Powerful Purpose Associates

"Anthony has been a valuable asset to our committee the past few years being a regular participant at our meetings and taking a lead role on some of our sub committees. He recently launched our LinkedIn YEAC page and helps maintain it. He is not afraid to speak up and voice his opinion even if different from the rest of the group and volunteers freely to assist on various tasks." *September 30, 2011*

(2nd) Carlos Gittens, P.E., *Chair, Nat'l Society of Prof. Engrs - Prof. Engrs in Private Practice - Young Engrs Advisory Council*
worked directly with Anthony at NSPE PEPP Young Engineer's Advisory Council

"I have had the pleasure of knowing Anthony for the past 5 years. Anthony and I have provided local seminars on career management and social networking topics. Anthony is a dynamic speaker who speaks with confidence and passion. He is honest and professional at all times. He has an outstanding background in the engineering field and is able to communicate his experience to others within and outside the industry." *January 21, 2011*

(2nd) Al Tretola, CPA, *Principal, The Gotham Search Group*
worked with Anthony at Powerful Purpose Associates

"Anthony is an expert in utilizing social media to leverage business contacts and create leads and sales. I attended Anthony's Seminar on LinkedIn last night and was so pumped up from his discussions and tips that I stayed up till 3 am to complete my profile. I went from 40% complete to 100% in less than 10 hours. Anthony's leadership and motivational skills are unbeatable and I highly recommend his coaching services to small businesses and professional groups." *November 23, 2010*

Top qualities: Great Results, Expert, High Integrity

(2nd) Bob Bauerle, P.E.
hired Anthony as a Business Consultant in 2010, and hired Anthony more than once

Figure 12-8 LinkedIn recommendations profile for Anthony Fasano.
Source: www.linkedin.com/in/anthonyjfasano

Connect with me on LinkedIn: www.linkedin.com/in/jillbrockmann

Visit the companion website to see step-by-step instructions for designing and implementing your very own LinkedIn account. (See the Preface for the URL.)

TARGETING EXPERTS WITH GOOGLE+

Have you ever heard the word "google" used as a verb? Usually people say, "Google it!" That is probably because Google is the largest search engine on the Internet. Wouldn't it be terrific if your engineering business came up at the top of the list when someone searched, "expert chemical engineer"?

Google+, Google Plus or just g+, is the brain-child of Google, Inc. Google is not just a search engine. It has developed many collaboration-based applications that are widely used throughout the engineering field. In an attempt to gather all of these cutting-edge applications into one place, Google engineers created Google+. This platform integrates many of Google's products to create a broader social experience. The result is a platform for professionals who want to increase their online visibility and personal brand identity.

Figure 12-9 is an example of the Google+ page for the engineering program at Stanford University.

Figure 12-9 Google+ profile page for Stanford University Engineering program.
Source: https://plus.google.com/u/0/102383602041872018960/posts

CREATING YOUR GOOGLE+ PROFILE PAGE

If you do not have a Google+ account yet, it is easy to create. Type the following text in your browser: http://plus.google.com

Google+ has an intuitive user interface that walks you through creating your profile in minutes. Visit the companion website to see step-by-step instructions for designing and customizing your own Google+ account. (See the Preface for the URL.)

CIRCLING ENGINEERING EXPERTS

The main functions of Google+ are bundled to include a search engine, an email client, an Internet browser, friend streams, circles of specialized contacts, group video chats with a new feature called Hangouts, personalized and automated search functions, the ability to target engineering industry experts, and the community-building feature called Circles.

The Circles feature of Google+, at first glance, may look the same as "friending" people on Facebook. However, Circles is a much more intricate way to organize lists (circles) of engineering industry experts, colleagues, and friends. The streamlining of these circles of contacts allows you to create entirely separate groups (circles) for family, friends, college alumni, colleagues, civil engineering experts, aeronautic engineers, mechanical engineers, sports fans, and more.

The main point to remember about Google+ Circles is that it allows you to read what you want, share what you want, share it when you want, and share it with whom you want. It also allows you to filter out the noise from everyone in your circles and drill down to reading only the content posted by the engineering experts you choose. This ability to focus on a specific topic is a huge time saver. For example, if you want to read about new developments in wind turbine grid engineering, you simply click the circle (that you created and filled with experts) named "wind turbine experts" and scroll through the posts of those experts.

Figure 12-10 shows an example of a Google+ Circles page.

Creating, adding, deleting, and modifying your Circles is as easy as a click or two. You can also invite people who are in your circles to participate in a live online session called a "Hangout." These hangouts can be recorded and replayed via YouTube.com. This function enables you to conduct online meetings with others. There is no more flying to meetings. You can sit in your home office in your pajamas! You can join other Hangouts as an active participant or simply observe and listen.

MAXIMIZING YOUR ONLINE VISIBILITY

A useful feature of using Google+ in an engineering business is that your Google+ page content is part of Google's web search integration and the ability for consumers to "direct connect" to your page. Through this connection ability, consumers or clients are directed to your website or other online venues, such as your blog.

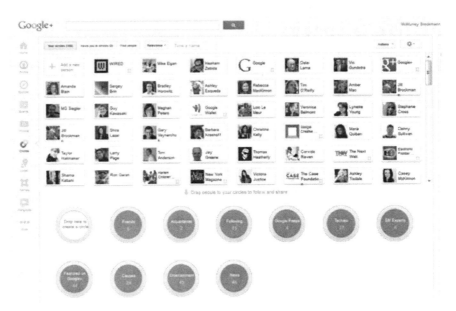

Figure 12-10 Google+ Circles page for McMurrey-Brockmann Educational Resources, LLC.
Source: https://plus.google.com/McBrockEdu

Here are some examples of what to think about when creating your Google+ account:

- How will you find and follow industry experts in your field of engineering?
- How will you distinguish yourself from other engineers on Google+?
- How will you gain more visibility and prove your expertise on Google+ by adding new connections?

Add me to your Circles on Google+ at: Jill@Get-ACE.com

Visit the companion website to see step-by-step instructions for designing and implementing your own Google+ account. (See the Preface for the URL.)

To "bug" or "debug"? That's the question.

In 1947, Harvard University engineering students found the first "computer bug." They taped the moth in their logbook and labeled it "first actual case of bug being found." The words "bug" and "debug" soon became mainstays in the language of computer programmers.

For details, see the Preface for the URL.

BIBLIOGRAPHY

Brad Friedman. *Twitter Facts and Stats*. http://socialmediatoday.com/bradfriedman/469107
/twitter-facts-and-stats. Accessed July 1, 2012.

Elad, Joel. *LinkedIn for Dummies*. Hoboken, NJ: John Wiley & Sons, 2011.

Gentle, Anne. *Conversation and Community: The Social Web for Documentation*. Fort Collins,
CO: XML Press, 2009.

Hay, Deltina. *The Social Media Survival Guide*. Fresno, CA: Quill Driver Books, 2011.

Kabani, Shama. *The Zen of Social Media Marketing*. Dallas, TX: Ben Bella Books, 2012.

Neuman-Beck, J. and Beck, M. *Visual Quickstart Guide for WordPress*. Berkeley, CA: Peachpit
Press, 2012.

Rowse, D. and Garrett, C. *ProBlogger*. Indianapolis, IN: John Wiley & Sons, 2012.

TEXTBOX SOURCES:

Umbrella: www.interestingengineering.com/2012/07/this-umbrella-will-charge-your
-phone.html

Stockholm: www.ivanhoe.com/science/story/2011/08/891a.html

Oresund Bridge: www.engineeringdaily.net/spotlight-project-the-oresund-bridge

Sound for quiet cars: www.ivanhoe.com/science/story/2010/11/786si.html

Algae-Powered Car: http://inhabitat.com/first-algae-powered-car-attempts-to-cross-us-on
-25-gallons

Computer Bug: http://americanhistory.si.edu/collections/comphist/objects/bug.htm

INDEX

CPSIA information can be obtained at www.ICGtesting.com
Printed in the USA
BVOW022008090513

320366BV00002B/15/P